"THIS YOU CAN'T MISS."
New York Herald Tribune

Here is a collection of outstanding stories created by the most imaginative writers of our day, selected from THE BEST SCIENCE FICTION STORIES, the annual anthology which the *St. Louis Post-Dispatch* calls "the cream of Science Fiction."

"Gems set the standards in this collection..."
New York Times

FRONTIERS IN SPACE

EDITED BY
EVERETT F. BLEILER
AND
T. E. DIKTY

SELECTIONS FROM
THE BEST SCIENCE FICTION STORIES

BANTAM BOOKS   NEW YORK

FRONTIERS IN SPACE

Selections from
The Best Science-Fiction Stories—1951, 1952 and 1953

A BANTAM Book published by arrangement with
Frederick Fell, Inc.

Bantam Edition published May, 1955
1st Printing............April, 1955

Copyright, 1951, 1952, 1953 by Everett F. Bleiler
and T. E. Dikty.

The following stories are reprinted from *The Best Science-Fiction Stories, 1951*, published by Frederick Fell, Inc.: ODDY AND ID, PROCESS, THE STAR DUCKS, TO SERVE MAN, THE FOX IN THE FOREST.

The following stories are reprinted from *The Best Science-Fiction Stories, 1952*, published by Frederick Fell, Inc.: NINE-FINGER JACK, DARK INTERLUDE, GENERATION OF NOAH, THE RATS.

The following stories are reprinted from *The Best Science-Fiction Stories, 1953*, published by Frederick Fell, Inc.: ARARAT, THE MOON IS GREEN, SURVIVAL, MACHINE, I AM NOTHING.

All rights in this book are reserved. It may not be used for dramatic, motion, or talking-picture purposes without written authorization from the holder of these rights. Nor may the book or any part thereof be reproduced in any manner whatever without permission in writing, except for brief quotations embodied in critical articles and reviews. For information, address: FREDERICK FELL, INC., 386 Fourth Avenue, New York 16, N. Y.

Bantam Books are published by Bantam Books, Inc. Its trade mark, consisting of the words "BANTAM BOOKS" and the portrayal of a bantam, is registered in the U. S. Patent Office and in other countries. *Marca Registrada*

PRINTED IN THE UNITED STATES OF AMERICA
BANTAM BOOKS, 25 West 45th Street, New York 36, N. Y.

CONTENTS

ODDY AND ID, Alfred Bester	1
PROCESS, A. E. Van Vogt	15
THE STAR DUCKS, Bill Brown	21
TO SERVE MAN, Damon Knight	29
THE FOX IN THE FOREST, Ray Bradbury	37
NINE-FINGER JACK, Anthony Boucher	51
DARK INTERLUDE, Mack Reynolds and Fredric Brown	56
GENERATION OF NOAH, William Tenn	63
THE RATS, Arthur Porges	74
ARARAT, Zenna Henderson	83
THE MOON IS GREEN, Fritz Leiber	105
SURVIVAL, John Wyndham	120
MACHINE, John W. Jakes	144
I AM NOTHING, Eric Frank Russell	148

ODDY AND ID

By Alfred Bester

THIS is the story of a monster.

They named him Odysseus Gaul in honor of Papa's favorite hero, and over Mama's desperate objections; but he was known as Oddy from the age of one.

The first year of life is an egotistic craving for warmth and security. Oddy was not likely to have much of that when he was born, for Papa's real estate business was bankrupt, and Mama was thinking of divorce. But an unexpected decision by United Radiation to build a plant in the town made Papa wealthy, and Mama fell in love with him all over again. So Oddy had warmth and security.

The second year of life is a timid exploration. Oddy crawled and explored. When he reached for the crimson coils inside the non-objective fireplace, an unexpected short-circuit saved him from a burn. When he fell out the third floor window, it was into the grass filled hopper of the Mechano-Gardener. When he teased the Phoebus Cat, it slipped as it snapped at his face, and the brilliant fangs clicked harmlessly over his ear.

"Animals love Oddy," Mama said. "They only pretend to bite."

Oddy wanted to be loved, so everybody loved Oddy. He was petted, pampered and spoiled through pre-school age. Shopkeepers presented him with largess, and acquaintances showered him with gifts. Of sodas, candy, tarts, chrystons, bobbletucks, freezies and various other comestibles, Oddy consumed enough for an entire kindergarten. He was never sick.

"Takes after his father," Papa said. "Good stock."

Family legends grew about Oddy's luck. . . . How a perfect stranger mistook him for his own child just as Oddy was about to amble into the Electronic Circus, and delayed him long enough to save him from the disastrous explosion of '98. . . . How a forgotten library book rescued him from the Rocket

Copyright 1950 by Street and Smith Publications, Inc., and originally published in *Astounding Science Fiction*, August, 1950.

Crash of '99. . . . How a multitude of odd incidents saved him from a multitude of assorted catastrophes. No one realized he was a monster . . . yet.

At eighteen, he was a nice looking boy with seal brown hair, warm brown eyes and a wide grin that showed even white teeth. He was strong, healthy, intelligent. He was completely uninhibited in his quiet, relaxed way. He had charm. He was happy. So far, his monstrous evil had only affected the little Town Unit where he was born and raised.

He came to Harvard from a Progressive School, so when one of his many quick friends popped into the dormitory room and said: "Hey Oddy, come down to the Quad and kick a ball around," Oddy answered: "I don't know how, Ben."

"Don't know how?" Ben tucked the football under his arm and dragged Oddy with him. "Where you been, laddie?"

"They didn't talk much about football back home," Oddy grinned. "Thought it was old fashioned. We were strictly Huxley-Hob."

"Huxley-Hob! That's for hi-brows," Ben said. "Football is still the big game. You want to be famous? You got to be on that gridiron before the Video every Saturday."

"So I've noticed, Ben. Show me."

Ben showed Oddy, carefully and with patience. Oddy took the lesson seriously and industriously. His third punt was caught by a freakish gust of wind, travelled seventy yards through the air, and burst through the third floor window of Proctor Charley (Gravy-Train) Stuart. Stuart took one look out the window and had Oddy down to Soldier Stadium in half an hour. Three Saturdays later, the headlines read: ODDY GAUL 57–ARMY 0.

"Snell & Rumination!" Coach Hig Clayton swore. "How does he do it? There's nothing sensational about that kid. He's just average. But when he runs they fall down chasing him. When he kicks, they fumble. When they fumble, he recovers."

"He's a negative player," Gravy-Train answered. "He lets you make the mistakes and then he cashes in."

They were both wrong. Oddy Gaul was a monster.

With his choice of any eligible young woman, Oddy Gaul went stag to the Observatory Prom, wandered into a darkroom by mistake, and discovered a girl in a smock bending over trays in the hideous green safe-light. She had cropped black hair, icy blue eyes, strong features, and a sensuous boyish figure. She ordered him out and Oddy fell in love with her . . . temporarily.

His friends howled with laughter when he told them.

"Shades of Pygmalion, Oddy, don't you know about *her?* The girl is frigid. A statue. She loathes men. You're wasting your time."

But through the adroitness of her analyst, the girl turned a neurotic corner one week later and fell deeply in love with Oddy Gaul. It was sudden, devastating and enraptured for two months. Then just as Oddy began to cool, the girl had a relapse and everything ended on a friendly, convenient basis.

So far only minor events made up the response to Oddy's luck, but the shock-wave reaction was spreading. In September of his Sophomore year, Oddy competed for the Political Economy Medal with a thesis entitled: "Causes of Mutiny." The striking similarity of his paper to the Astraean Mutiny that broke out the day his paper was entered won him the prize.

In October, Oddy contributed twenty dollars to a pool organized by a crack-pot classmate for speculating on the Exchange according to "Stock Market Trends," a thousand year old superstition. The seer's calculations were ridiculous, but a sharp panic nearly ruined the Exchange as it quadrupled the pool. Oddy made one hundred dollars.

And so it went . . . worse and worse. The monster.

Now a monster can get away with a lot when he's studying speculative philosophy where causation is rooted in history and the Present is devoted to statistical analysis of the Past; but the living sciences are bulldogs with their teeth clamped on the phenomena of Now. So it was Jesse Migg, physiologist and spectral physicist, who first trapped the monster . . . and he thought he was an angel.

Old Jess was one of the Sights. In the first place he was young . . . not over forty. He was a malignant knife of a man, an albino, pink-eyed, bald, pointed-nosed and brilliant. He affected 20th Century clothes and 20th Century vices . . . tobacco and potations of C_2H_5OH. He never talked . . . He spat. He never walked . . . He scurried. And he was scurrying up and down the aisles of the laboratory of Tech I (General Survey of Spatial Mechanics—Required for All General Arts Students) when he ferreted out the monster.

One of the first experiments in the course was EMF Electrolysis. Elementary stuff. A U-Tube containing water was passed between the poles of a stock Remosant Magnet. After sufficient voltage was transmitted through the coils, you drew off Hydrogen and Oxygen in two-to-one ratio at the arms of the tube and related them to the voltage and the magnetic field.

Oddy ran his experiment earnestly, got the proper results,

entered them in his lab book and then waited for the official check-off. Little Migg came hustling down the aisle, darted to Oddy and spat: "Finished?"

"Yes, sir."

Migg checked the book entries, glanced at the indicators at the ends of the tube, and stamped Oddy out with a sneer. It was only after Oddy was gone that he noticed the Remosant Magnet was obviously shorted. The wires were fused. There hadn't been any field to electrolyse the water.

"Curse and Confusion!" Migg grunted (he also affected 20th Century vituperation) and rolled a clumsy cigarette.

He checked off possibilities in his comptometer head. 1. Gaul cheated. 2. If so, with what apparatus did he portion out the H_2 and O_2? 3. Where did he get the pure gases? 4. Why did he do it? Honesty was easier. 5. He didn't cheat. 6. How did he get the right results? 7. How did he get *any* results?

Old Jess emptied the U-Tube, refilled it with water and ran off the experiment himself. He too got the correct result without a magnet.

"Rice on a Raft!" he swore, unimpressed by the miracle, and infuriated by the mystery. He snooped, darting about like a hungry bat. After four hours he discovered that the steel bench supports were picking up a charge from the Greeson Coils in the basement and had thrown just enough field to make everything come out right.

"Coincidence," Migg spat. But he was not convinced.

Two weeks later, in Elementary Fission Analysis, Oddy completed his afternoon's work with a careful listing of resultant isotopes from selenium to lanthanum. The only trouble, Migg discovered, was that there had been a mistake in the stock issued to Oddy. He hadn't received any U^{235} for neutron bombardment. His sample had been a left-over from a Stefan-Boltzmann black-body demonstration.

"Frog in Heaven!" Migg swore, and double-checked. Then he triple-checked. When he found the answer . . . a remarkable coincidence involving improperly cleaned apparatus and a defective cloud-chamber, he swore further. He also did some intensive thinking.

"There are accident prones," Migg snarled at the reflection in his Self-Analysis Mirror. "How about Good Luck prones? Horse Manure!"

But he was a bulldog with his teeth sunk in phenomena. He tested Oddy Gaul. He hovered over him in the laboratory, cackling with infuriated glee as Oddy completed experiment after experiment with defective equipment. When Oddy successfully completed the Rutherford Classic . . . getting $_8O^{17}$ after exposing nitrogen to alpha radiation . . . but in this case without the use of nitrogen or alpha radiation, Migg actually clapped him on the back in delight. Then the little man investigated and found the logical, improbable chain of coincidences that explained it.

He devoted his spare time to a check-back on Oddy's career at Harvard. He had a two hour conference with a lady astronomer's faculty analyst, and a ten minute talk with Hig Clayton and Gravy-Train Stuart. He rooted out the Exchange Pool, the Political Economy Medal, and half a dozen other incidents that filled him with malignant joy. Then he cast off his 20th Century affectation, dressed himself properly in formal leotards, and entered the Faculty Club for the first time in a year.

A four-handed chess game in three dimensions was in progress in the Diathermy Alcove. It had been in progress since Migg joined the faculty, and would probably not be finished before the end of the century. In fact, Johansen, playing Red, was already training his son to replace him in the likely event of his dying before the completion of the game.

As abrupt as ever, Migg marched up to the glowing cube, sparkling with sixteen layers of vari-colored pieces, and blurted: "What do you know about accidents?"

"Ah?" said Bellanby, *Philosopher in Res* at the University. "Good evening, Migg. Do you mean the accident of substance, or the accident of essence? If, on the other hand, your question implies—"

"No, no," Migg interrupted. "My apologies, Bellanby. Let me rephrase the question. Is there such a thing as Compulsion of Probability?"

Hrrdnikkisch completed his move and gave full attention to Migg, as did Johansen and Bellanby. Wilson continued to study the board. Since he was permitted one hour to make his move and would need it, Migg knew there would be ample time for the discussion.

"Compulthon of Probability?" Hrrdnikkisch lisped. "Not a new conthept, Migg. I recall a thurvey of the theme in 'The Integraph' Vol. LVIII, No. 9. The calculuth, if I am not mithtaken—"

"No," Migg interrupted again. "My respects, Signoid. I'm not interested in the mathematic of Probability, nor the philosophy. Let me put it this way. The Accident Prone has already been incorporated into the body of Psychoanalysis. Paton's Theorem of the Least Neurotic Norm settled that. But

I've discovered the obverse. I've discovered a Fortune Prone."

"Ah?" Johansen chuckled. "It's to be a joke. You wait and see, Signoid."

"No," answered Migg. "I'm perfectly serious. I've discovered a genuinely lucky man."

"He wins at cards?"

"He wins at everything. Accept this postulate for the moment . . . I'll document it later . . . There is a man who is lucky. He is a Fortune Prone. Whatever he desires, he receives. Whether he has the ability to achieve it or not, he receives it. If his desire is totally beyond the peak of his accomplishment, then the factors of chance, coincidence, hazard, accident . . . and so on, combine to produce his desired end."

"No." Bellanby shook his head. "Too far-fetched."

"I've worked it out empirically," Migg continued. "It's something like this. The future is a choice of mutually exclusive possibilities, one or other of which must be realized in terms of favorability of the events and number of the events . . ."

"Yes, yes," interrupted Johansen. "The greater the number of favorable possibilities, the stronger the probability of an event maturing. This is elementary, Migg. Go on."

"I continue," Migg spat indignantly. "When we discuss Probability in terms of throwing dice, the predictions or odds are simple. There are only six mutually exclusive possibilities to each die. The favorability is easy to compute. Chance is induced to simple odds-ratios. *But* when we discuss probability in terms of the Universe, we cannot encompass enough data to make a prediction. There are too many factors. Favorability cannot be ascertained."

"All thith ith true," Hrrdnikkisch said, "but what of your Fortune Prone?"

"I don't know how he does it . . . but merely by the intensity or mere existence of his desire, he can affect the favorability of possibilities. By wanting, he can turn possibility into probability, and probability into certainty."

"Ridiculous," Bellanby snapped. "You claim there's a man far-sighted and far-reaching enough to do this?"

"Nothing of the sort. He doesn't know what he's doing. He just thinks he's lucky, if he thinks about it at all. Let us say he wants . . . Oh . . . Name anything."

"Heroin," Bellanby said.

"What's that?" Johansen inquired.

"A morphine derivative," Hrrdnikkisch explained. "Formerly manufactured and thold to narcotic addictth."

Oddy and Id

"Heroin," Migg said. "Excellent. Say my man desires Heroin, an antique narcotic no longer in existence. Very good. His desire would compel this sequence of possible but improbable events: A chemist in Australia, fumbling through a new organic synthesis, will accidentally and unwittingly prepare six ounces of Heroin. Four ounces will be discarded, but through a logical mistake two ounces will be preserved. A further coincidence will ship it to this country and this city, wrapped as powdered sugar in a plastic ball; where the final accident will serve it to my man in a restaurant which he is visiting for the first time on an impulse. . . ."

"La-La-La!" said Hrrdnikkisch. "Thith shuffling of hithory. Thith fluctuation of inthident and pothibility? All achieved without the knowledge but with the dethire of a man?"

"Yes. Precisely my point," Migg snarled. "I don't know how he does it, but he turns possibility into certainty. And since almost anything is possible, he is capable of accomplishing almost anything. He is God-like but not a God because he does this without consciousness. He is an angel."

"Who is this angel?" Johansen asked.

And Migg told them all about Oddy Gaul.

"But how does he do it?" Bellanby persisted. "How does he do it?"

"I don't know," Migg repeated again. "Tell me how Espers do it."

"What!" Bellanby exclaimed. "Are you prepared to deny the EK pattern of thought? Do you—"

"I do nothing of the sort. I merely illustrate one possible explanation. Man produces events. The threatening War of Resources may be thought to be a result of the natural exhaustion of terran resources. We know it is not. It is a result of centuries of thriftless waste by man. Natural phenomena are less often produced by nature and more often produced by man."

"And?"

"Who knows? Gaul is producing phenomena. Perhaps he's unconsciously broadcasting on an EK waveband. Broadcasting and getting results. He wants Heroin. The broadcast goes out—"

"But Espers can't pick up any EK brain pattern further than the horizon. It's direct wave transmission. Even large objects cannot be penetrated. A building, say, or a—"

"I'm not saying this is on the Esper level," Migg shouted. "I'm trying to imagine something bigger. Something tremendous. He wants Heroin. His broadcast goes out to the world.

All men unconsciously fall into a pattern of activity which will produce that Heroin as quickly as possible. That Austrian chemist—"

"No. Australian."

"That Australian chemist may have been debating between half a dozen different syntheses. Five of them could never have produced Heroin; but Gaul's impulse made him select the sixth."

"And if he did not anyway?"

"Then who knows what parallel chains were also started? A boy playing Cops and Robbers in Montreal is impelled to explore an abandoned cabin where he finds the drug, hidden there centuries ago by smugglers. A woman in California collects old apothecary jars. She finds a pound of Heroin. A child in Berlin, playing with a defective Radar-Chem Set, manufactures it. Name the most improbable sequence of events, and Gaul can bring it about, logically and certainly. I tell you, that boy is an angel!"

And he produced his documented evidence and convinced them.

It was then that four scholars of various but indisputable intellects elected themselves an executive committee for Fate and took Oddy Gaul in hand. To understand what they attempted to do, you must first understand the situation the world found itself in during that particular era.

It is a known fact that all wars are founded in economic conflict, or to put it another way, a trial by arms is merely the last battle of an economic war. In the pre-Christian centuries, the Punic Wars were the final outcome of a financial struggle between Rome and Carthage for economic control of the Mediterranean. Three thousand years later, the impending War of Resources loomed as the finale of a struggle between the two Independent Welfare States controlling most of the known economic world.

What petroleum oil was to the 20th Century, FO (the nickname for Fissionable Ore) was to the 30th; and the situation was peculiarly similar to the Asia Minor crisis that ultimately wrecked the United Nations a thousand years before. Triton, a backward, semibarbaric satellite, previously unwanted and ignored, had suddenly discovered it possessed enormous resources of FO. Financially and technologically incapable of self-development, Triton was peddling concessions to both Welfare States.

The difference between a Welfare State and a Benevolent Despot is slight. In times of crisis, either can be traduced by the sincerest motives into the most abominable conduct. Both the Comity of Nations (bitterly nicknamed "The Con Men" by Der Realpolitik aus Terra) and Der Realpolitik aus Terra (sardonically called "The Rats" by the Comity of Nations) were desperately in need of natural resources, meaning FO. They were bidding against each other hysterically, and elbowing each other with sharp skirmishes at outposts. Their sole concern was the protection of their citizens. From the best of motives they were preparing to cut each other's throat.

Had this been the issue before the citizens of both Welfare States, some compromise might have been reached; but Triton in the catbird seat, intoxicated as a schoolboy with newfound prominence and power, confused issues by raising a religious question and reviving a Holy War which the Family of Planets had long forgotten. Assistance in their Holy War (involving the extermination of a harmless and rather unimportant sect called the Quakers) was one of the conditions of sale. This, both the Comity of Nations and Der Realpolitik aus Terra were prepared to swallow with or without private reservations, but it could not be admitted to their citizens.

And so, camouflaged by the burning issues of Rights of Minority Sects, Priority of Pioneering, Freedom of Religion, Historical Rights to Triton v. Possession in Fact, etc., the two Houses of the Family of Planets feinted, parried, riposted and slowly closed, like fencers on the strip, for the final sortie which meant ruin for both.

All this the four men discussed through three interminable meetings.

"Look here," Migg complained toward the close of the third consultation. "You theoreticians have already turned nine man-hours into carbonic acid with ridiculous dissensions . . ."

Bellanby nodded, smiling. "It's as I've always said, Migg. Every man nurses the secret belief that were he God he could do the job much better. We're just learning how difficult it is."

"Not God," Hrrdnikkisch said, "but hith Prime Minithterth. Gaul will be God."

Johansen winced. "I don't like that talk," he said. "I happen to be a religious man."

"You?" Bellanby exclaimed in surprise. "A Colloid-Therapeutist?"

"I happen to be a religious man," Johansen repeated stubbornly.

"But the boy hath the power of the miracle," Hrrdnikkisch protested. "When he hath been taught to know what he doeth, he will be a God."

"This is pointless," Migg rapped out. "We have spent three sessions in piffling discussion. I have heard three opposed views re Mr. Odysseus Gaul. Although all are agreed he must be used as a tool, none can agree on the work to which the tool must be set. Bellanby prattles about an Ideal Intellectual Anarchy, Johansen preaches about a Soviet of God, and Hrrdnikkisch has wasted two hours postulating and destroying his own theorems . . ."

"Really, Migg . . ." Hrrdnikkisch began. Migg waved his hand.

"Permit me," Migg continued malevolently, "to reduce this discussion to the kindergarten level. First things first, gentlemen. Before attempting to reach cosmic agreement we must make sure there is a cosmos left for us to agree upon. I refer to the impending war . . .

"Our program, as I see it, must be simple and direct. It is the education of a God or, if Johansen protests, of an angel. Fortunately Gaul is an estimable young man of kindly, honest disposition. I shudder to think what he might have done had he been inherently vicious."

"Or what he might do once he learns what he can do," muttered Bellanby.

"Precisely. We must begin a careful and rigorous ethical educcation of the boy, but we haven't enough time. We can't educate first, and then explain the truth when he's safe. We must forestall the war. We need a short-cut."

"All right," Johansen said. "What do you suggest?"

"Dazzlement," Migg spat. "Enchantment."

"Enchantment?" Hrrdnikkisch chuckled. "A new thienth, Migg?"

"Why do you think I selected you three of all people for this secret?" Migg snorted. "For your intellects? Nonsense! I can think you all under the table. No. I selected you, gentlemen, for your charm."

"It's an insult," Bellanby grinned, "and yet I'm flattered."

"Gaul is nineteen," Migg went on. "He is at the age when undergraduates are most susceptible to hero-worship. I want you gentlemen to charm him. You are not the first brains of the University, but you are the first heroes."

"I altho am inthulted and flattered," said Hrrdnikkisch.

"I want you to charm him, dazzle him, inspire him with affection and awe . . . as you've done with countless classes of undergraduates."

"Aha!" said Johansen. "The chocolate around the pill."

"Exactly. When he's enchanted, you will make him want to

stop the war . . . and then tell him how he can stop it. That will give us breathing space to continue his education. By the time he outgrows his respect for you he will have a sound ethical foundation on which to build. He'll be safe."

"And you, Migg?" Bellanby inquired. "What part do you play?"

"Now? None," Migg snarled. "I have no charm, gentlemen. I come later. When he outgrows his respect for you, he'll begin to acquire respect for me."

All of which was frightfully conceited but perfectly true.

And as events slowly marched toward the final crisis, Oddy Gaul was carefully and quickly enchanted. Bellanby invited him to the twenty foot crystal globe atop his house . . . the famous hen-roost to which only the favored few were invited. There, Oddy Gaul sunbathed and admired the philosopher's magnificent iron-hard condition at seventy-three. Admiring Bellanby's muscles, it was only natural for him to admire Bellanby's ideas. He returned often to sunbathe, worship the great man, and absorb ethical concepts.

Meanwhile, Hrrdnikkisch took over Oddy's evenings. With the mathematician, who puffed and lisped like some flamboyant character out of Rabelais, Oddy was carried to the dizzy heights of the *haute cuisine* and the complete pagan life. Together they ate and drank incredible foods and liquids and pursued incredible women until Oddy returned to his room each night, intoxicated with the magic of the senses and the riotous color of the great Hrrdnikkisch's glittering ideas.

And occasionally . . . not too often, he would find Papa Johansen waiting for him, and then would come the long quiet talks through the small hours when young men search for the harmonics of life and the meaning of entity. And there was Johansen for Oddy to model himself after . . . a glowing embodiment of Spiritual Good . . . a living example of Faith in God and Ethical Sanity.

The climax came on March 15th . . . The Ides of March, and they should have taken the date as a sign. After dinner with his three heroes at the Faculty Club, Oddy was ushered into the Foto-Library by the three great men where they were joined, quite casually, by Jesse Migg. There passed a few moments of uneasy tension until Migg made a sign, and Bellanby began.

"Oddy," he said, "have you ever had the fantasy that some day you might wake up and discover you were a King?"

Oddy blushed.

"I see you have. You know, every man has entertained that

dream. The usual pattern is: You learn your parents only adopted you, and that you are actually and rightfully the King of . . . of . . ."

"Baratraria," said Hrrdnikkisch who had made a study of Stone Age Fiction.

"Yes, sir," Oddy muttered. "I've had that dream."

"Well," Bellanby said quietly, "it's come true. You are a King."

Oddy stared while they explained and explained and explained. First, as a college boy, he was wary and suspicious of a joke. Then, as an idolator, he was almost persuaded by the men he most admired. And finally, as a human animal, he was swept away by the exaltation of security. Not power, not glory, not wealth thrilled him, but security alone. Later he might come to enjoy the trimmings, but now he was released from fear. He need never worry again.

"Yes," exclaimed Oddy. "Yes, yes, yes! I understand. I understand what you want me to do." He surged up excitedly from his chair and circled the illuminated walls, trembling with joy and intoxication. Then he stopped and turned.

"And I'm grateful," he said. "Grateful to all of you for what you've been trying to do. It would have been shameful if I'd been selfish . . . or mean . . . Trying to use this for myself. But you've shown me the way. It's to be used for good. Always!"

Johansen nodded happily.

"I'll always listen to you," Oddy went on. "I don't want to make any mistakes. Ever!" He paused and blushed again. "That dream about being a king . . . I had that when I was a kid. But here at the school I've had something bigger. I used to wonder what would happen if I was the one man who could run the world. I used to dream about the kind things I'd do . . ."

"Yes," said Bellanby. "We know, Oddy. We've all had that dream too. Every man does."

"But it isn't a dream any more," Oddy laughed. "It's reality. I can do it. I can make it happen."

"Start with the war," Migg said sourly.

"Of course," said Oddy. "The war first; but then we'll go on from there, won't we? I'll make sure the war never starts, but then we'll do big things . . . great things! Just the five of us in private. Nobody'll know about us. We'll be ordinary people, but we'll make life wonderful for everybody. If I'm an angel . . . like you say . . . then I'll spread heaven around me as far as I can reach."

Oddy and Id

"But start with the war," Migg repeated.

"The war is the first disaster that must be averted, Oddy," Bellanby said. "If you don't want this disaster to happen, it will never happen."

"And you want to prevent that tragedy, don't you?" said Johansen.

"Yes," said Oddy. "I do."

On March 20th, the war broke. The Comity of Nations and Der Realpolitik aus Terra mobilized and struck. While blow followed shattering counter-blow, Oddy Gaul was commissioned Subaltern in a Line regiment, but gazetted to Intelligence on May 3rd. On June 24th he was appointed A.D.C. to the Joint Forces Council meeting in the ruins of what had been Australia. On July 11th he was brevetted to command of the wrecked Space Force, being jumped 1,789 grades over regular officers. On September 19th he assumed supreme command in the Battle of the Parsec and won the victory that ended the disastrous solar annihilation called the Six Month War.

On September 23rd, Oddy Gaul made the astonishing Peace Offer that was accepted by the remnants of both Welfare States. It required the scrapping of antagonistic economic theories, and amounted to the virtual abandonment of all economic theory with an amalgamation of both States into a Solar Society. On January 1st, Oddy Gaul, by unanimous acclaim, was elected Solon of the Solar Society in perpetuity.

And today . . . still youthful, still vigorous, still handsome, still sincere, idealistic, charitable, kindly and sympathetic, he lives in the Solar Palace. He is unmarried but a mighty lover; uninhibited, but a charming host and devoted friend; democratic, but the feudal overlord of a bankrupt Family of Planets that suffers misgovernment, oppression, poverty and confusion with a cheerful joy that sings nothing but Hosannahs to the glory of Oddy Gaul.

In a last moment of clarity, Jesse Migg communicated his desolate summation of the situation to his friends in the Faculty Club. This was shortly before they made the trip to join Oddy in the palace as his confidential and valued advisers.

"We were fools," Migg said bitterly. "We should have killed him. He isn't an angel. He's a monster. Civilization and culture . . . philosophy and ethics . . . Those were only masks Oddy put on; masks that covered the primitive impulses of his subconscious mind."

"You mean Oddy was not sincere?" Johansen asked heavily. "He wanted this wreckage . . . this ruin?"

"Certainly he was sincere . . . consciously. He still is. He thinks he desires nothing but the most good for the most men. He's honest, kind and generous . . . but only consciously."

"Ah! The Id!" said Hrrdnikkisch with an explosion of breath as though he had been punched in the stomach.

"You understand, Signoid? I see you do. Gentlemen, we were imbeciles. We made the mistake of assuming that Oddy would have conscious control of his power. He does not. The control was and still is below the thinking, reasoning level. The control lies in Oddy's Id . . . in that deep unconscious reservoir of primordial selfishness that lies within every man."

"Then he wanted the war," Bellanby said.

"His Id wanted the war, Bellanby. It was the quickest route to what his Id desires . . . to be Lord of the Universe and Loved by the Universe . . . and his Id controls the Power. All of us have that selfish, egocentric Id within us, perpetually searching for satisfaction, timeless, immortal, knowing no logic, no values, no good and evil, no morality; and that is what controls the power in Oddy. He will always get not what he's been educated to desire but what his Id desires. It's the inescapable conflict that may be the doom of our system."

"But we'll be there to advise him . . . counsel him . . . guide him," Bellanby protested. "He asked us to come."

"And he'll listen to our advice like the good child that he is," Migg answered, "agreeing with us, trying to make a heaven for everybody while his Id will be making a hell for everybody. Oddy isn't unique. We all suffer from the same conflict . . . but Oddy has the power."

"What can we do?" Johansen groaned. "What can we do?"

"I don't know." Migg bit his lip, then bobbed his head to Papa Johansen in what amounted to apology for him. "Johansen," he said, "you were right. There must be a God, if only because there must be an opposite to Oddy Gaul who was most assuredly invented by the Devil."

But that was Jesse Migg's last sane statement. Now, of course, he adores Gaul the Glorious, Gaul the Gauleiter, Gaul the God Eternal who has achieved the savage, selfish satisfaction for which all of us unconsciously yearn from birth, but which only Oddy Gaul has won.

PROCESS

By A. E. Van Vogt

IN THE bright light of that far sun, the forest breathed and had its being. It was aware of the ship that had come down through the thin mists of the upper air. But its automatic hostility to the alien things was not immediately accompanied by alarm.

For tens of thousands of square miles, its roots entwined under the ground, and its millions of tree tops swayed gently in a thousand idle breezes. And beyond, spreading over the hills and the mountains, and along almost endless sea coast were other forests as strong and as powerful as itself.

From time immemorial the forest had guarded the land from a dimly understood danger. What that danger was it began now slowly to remember. It was from ships like this, that descended from the sky. The forest could not recall clearly how it had defended itself in the past, but it did remember tensely that defense had been necessary.

Even as it grew more and more aware of the ship coasting along in the gray-red sky above, its leaves whispered a timeless tale of battles fought and won. Thoughts flowed their slow course down the channels of vibration, and the stately limbs of tens of thousands of trees trembled ever so slightly.

The vastness of that tremor, affecting as it did all the trees, gradually created a sound and a pressure. At first it was almost impalpable, like a breeze wafting through an evergreen glen. But it grew stronger.

It acquired substance. The sound became all-enveloping. And the whole forest stood there vibrating its hostility, waiting for the thing in the sky to come nearer.

It had not long to wait.

The ship swung down from its lane. Its speed, now that it was close to the ground, was greater than it had first seemed. And it was bigger. It loomed gigantic over the near trees, and

Copyright 1950 by Fantasy House, Inc., and originally published in *The Magazine of Fantasy and Science Fiction*, December, 1950.

swung down lower, careless of the tree tops. Brush crackled, limbs broke, and entire trees were brushed aside as if they were meaningless and weightless and without strength.

Down came the ship, cutting its own path through a forest that groaned and shrieked with its passage. It settled heavily into the ground two miles after it first touched a tree. Behind, the swath of broken trees quivered and pulsed in the light of the sun, a straight path of destruction which—the forest suddenly remembered—was exactly what had happened in the past.

It began to pull clear of the anguished parts. It drew out its juices, and ceased vibrating in the affected areas. Later, it would send new growth to replace what had been destroyed, but now it accepted the partial death it had suffered. It knew fear.

It was a fear tinged with anger. It felt the ship lying on crushed trees, on a part of itself that was not yet dead. It felt the coldness and the hardness of steel walls, and the fear and the anger increased.

A whisper of thought pulsed along the vibration channels. Wait, it said, there is a memory in me. A memory of long ago when other such ships as this came.

The memory refused to clarify. Tense but uncertain, the forest prepared to make its first attack. It began to grow around the ship.

Long ago it had discovered the power of growth that was possible to it. There was a time when it had not been as large as it was now. And then, one day, it became aware that it was coming near another forest like itself.

The two masses of growing wood, the two colossi of intertwined roots approached each other warily, slowly, in amazement, in a startled but cautious wonder that a similar life form should actually have existed all this time. Approached, touched —and fought for years.

During that prolonged struggle nearly all growth in the central portions stopped. Trees ceased to develop new branches. The leaves, by necessity, grew hardier, and performed their functions for much longer periods. Roots developed slowly. The entire available strength of the forest was concentrated in the processes of defense and attack.

Walls of trees sprang up overnight. Enormous roots tunneled into the ground for miles straight down, breaking through rock and metal, building a barrier of living wood against the encroaching growth of the strange forest. On the surface, the barriers thickened to a mile or more of trees that stood almost bole

Process

to bole. And, on that basis, the great battle finally petered out. The forest accepted the obstacle created by its enemy.

Later, it fought to a similar standstill a second forest which attacked it from another direction.

The limits of demarcation became as natural as the great salt sea to the south, or the icy cold of mountain tops that were frozen the year round.

As it had in battle with the two other forests, *the* forest concentrated its entire strength against the encroaching ship. Trees shot up at the rate of a foot every few minutes. Creepers climbed the trees, and flung themselves over the top of the vessel. The countless strands of it raced over the metal, and then twined themselves around the trees on the far side. The roots of those trees dug deeper into the ground, and anchored in rock strata heavier than any ship ever built. The tree boles thickened, and the creepers widened till they were enormous cables.

As the light of that first day faded into twilight, the ship was buried under thousands of tons of wood, and hidden in foliage so thick that nothing of it was visible.

The time had come for the final destructive action.

Shortly after dark, tiny roots began to fumble over the underside of the ship. They were infinitesimally small; so small that in the initial stages they were no more than a few dozens of atoms in diameter; so small that the apparently solid metal seemed almost emptiness to them; so incredibly small that they penetrated the hard steel effortlessly.

It was at that time, almost as if it had been waiting for this stage, that the ship took counteraction. The metal grew warm, then hot, and then cherry red. That was all that was needed. The tiny roots shriveled, and died. The larger roots near the metal burned slowly as the searing heat reached them.

Above the surface, other violence began. Flame darted from a hundred orifices of the ship's surface. First the creepers, then the trees began to burn. It was no flare-up of uncontrollable fire, no fierce conflagration leaping from tree to tree in irresistible fury. Long ago, the forest had learned to control fires started by lightning or spontaneous combustion. It was a matter of sending sap to the affected area. The greener the tree, the more sap that permeated it, then the hotter the fire would have to be.

The forest could not immediately remember ever having encountered a fire that could make inroads against a line of trees that oozed a sticky wetness from every crevice of their bark.

But this fire could. It was different. It was not only flame; it was energy. It did not feed off the wood; it was fed by an energy within itself.

That fact at last brought the associational memory to the forest. It was a sharp and unmistakable remembrance of what it had done long ago to rid itself and its planet of a ship just like this.

It began to withdraw from the vicinity of the ship. It abandoned the framework of wood and shrubbery with which it had sought to imprison the alien structure. As the precious sap was sucked back into trees that would now form a second line of defense, the flames grew brighter, and the fire waxed so brilliant that the whole scene was bathed in an eerie glow.

It was some time before the forest realized that the fire beams were no longer flaming out from the ship, and that what incandescence and smoke remained came from normally burning wood.

That, too, was according to its memory of what had happened—before.

Frantically though reluctantly the forest initiated what it now realized was the only method of ridding itself of the intruder. Frantically because it was hideously aware that the flame from the ship could destroy entire forests. And reluctantly because the method of defense involved its suffering the burns of energy only slightly less violent than those that had flared from the machine.

Tens of thousands of roots grew toward rock and soil formations that they had carefully avoided since the last ship had come. In spite of the need for haste, the process itself was slow. Tiny roots, quivering with unpleasant anticipation, forced themselves into the remote, buried ore beds, and by an intricate process of osmosis drews grains of pure metal from the impure natural stuff. The grains were almost as small as the roots that had earlier penetrated the steel walls of the ship, small enough to be borne along, suspended in sap, through a maze of larger roots.

Soon there were thousands of grains moving along the channels, then millions. And, though each was tiny in itself, the soil where they were discharged soon sparkled in the light of the dying fire. As the sun of that world reared up over the horizon, the silvery gleam showed a hundred feet wide all around the ship.

It was shortly after noon that the machine showed awareness of what was happening. A dozen hatches opened, and objects floated out of them. They came down to the ground, and began to skim up the silvery stuff with nozzled things that sucked up the fine dust in a steady fashion. They worked with great caution; but an hour before darkness set in again, they had scooped up more than twelve tons of the thinly spread Uranium 235.

As night fell, all the two-legged things vanished inside the vessel. The hatches closed. The long torpedo-shape floated lightly upward, and sped to the higher heavens where the sun still shone.

The first awareness of the situation came to the forest as the roots deep under the ship reported a sudden lessening of pressure. It was several hours before it decided that the enemy had actually been driven off. And several more hours went by before it realized that the uranium dust still on the scene would have to be removed. The rays spread too far afield.

The accident that occurred then took place for a very simple reason. The forest had taken the radioactive substance out of rock. To get rid of it, it need merely to put it back into the nearest rock beds, particularly the kind of rock that absorbed the radioactivity. To the forest the situation seemed as obvious as that.

An hour after it began to carry out the plan, the explosion mushroomed toward outer space.

It was vast beyond all the capacity of the forest to understand. It neither saw nor heard that colossal shape of death. What it did experience was enough. A hurricane leveled square miles of trees. The blast of heat and radiation started fires that took hours to put out.

Fear departed slowly, as it remembered that this too had happened before. Sharper by far than the memory was the vision of the possibilities of what had happened . . . the nature of the opportunity.

Shortly after dawn the following morning, it launched its attack. Its victim was the forest which—according to its faulty recollection—had originally invaded its territory.

Along the entire front which separated the two colossi, small atomic explosions erupted. The solid barrier of trees which was the other forest's outer defense went down before blast after blast of irresistible energy.

The enemy, reacting normally, brought up its reserve of sap. When it was fully committed to the gigantic task of growing a new barrier, the bombs started to go off again. The resulting explosions destroyed its main sap supply. And, since it did not understand what was happening, it was lost from that moment.

Into the no-man's-land where the bombs had gone off, the attacking forest rushed an endless supply of roots. Wherever resistance built up, there an atomic bomb went off. Shortly after the next noon, a titanic explosion destroyed the sensitive central trees—and the battle was over.

It took months for the forest to grow into the territory of its defeated enemy, to squeeze out the other's dying roots, to nudge over trees that now had no defense, and to put itself into full and unchallenged possession.

The moment the task was completed, it turned like a fury upon the forest on its other flank. Once more it attacked with atomic thunder, and with a hail of fire tried to overwhelm its opponent.

It was met by equal force. Exploding atoms!

For its knowledge had leaked across the barrier of intertwined roots which separated forests.

Almost, the two monsters destroyed each other. Each became a remnant, that started the painful process of regrowth. As the years passed, the memory of what had happened grew dim. Not that it mattered. Actually, the ships came at will. And somehow, even if the forest remembered, its atomic bombs would not go off in the presence of a ship.

The only thing that would drive away the ships was to surround each machine with a fine dust of radioactive stuff. Whereupon it would scoop up the material, and then hastily retreat.

Victory was always as simple as that.

THE STAR DUCKS

By Bill Brown

WARD RAFFERTY's long, sensitive newshawk's nose alerted him for a hoax as soon as he saw the old Alsop place. There was no crowd of curious farmers standing around, no ambulance.

Rafferty left *The Times* press car under a walnut tree in the drive and stood for a moment noting every detail with the efficiency that made him *The Times'* top reporter. The old Alsop house was brown, weathered, two-story with cream-colored filigree around the windows and a lawn that had grown up to weeds. Out in back were the barn and chicken houses and fences that were propped up with boards and pieces of pipe. The front gate was hanging by one hinge but it could be opened by lifting it. Rafferty went in and climbed the steps, careful for loose boards.

Mr. Alsop came out on the porch to meet him. "Howdy do," he said.

Rafferty pushed his hat back on his head the way he always did before he said: "I'm Rafferty of *The Times*." Most people knew his by-line and he liked to watch their faces when he said it.

"Rafferty?" Mr. Alsop said, and Rafferty knew he wasn't a *Times* reader.

"I'm a reporter," Rafferty said. "Somebody phoned in and said an airplane cracked up around here."

Mr. Alsop looked thoughtful and shook his head slowly.

"No," he said.

Rafferty saw right away that Alsop was a slow thinker so he gave him time, mentally pegging him a taciturn Yankee. Mr. Alsop answered again, "Noooooooooooo."

The screen door squeaked and Mrs. Alsop came out. Since Mr. Alsop was still thinking, Rafferty repeated the information for Mrs. Alsop, thinking she looked a little brighter than her

Copyright 1950 by Fantasy House, Inc., and originally published in *The Magazine of Fantasy and Science Fiction*, Fall, 1950.

husband. But Mrs. Alsop shook her head and said, "Nooooooooooo," in exactly the same tone Mr. Alsop had used.

Rafferty turned around with his hand on the porch railing ready to go down the steps.

"I guess it was just a phony tip," he said. "We get lots of them. Somebody said an airplane came down in your field this morning, straight down trailing fire."

Mrs. Alsop's face lighted up. "Ohhhhhhhhhh!" she said. "Yes it did but it wasn't wrecked. Besides, it isn't really an airplane. That is, it doesn't have wings on it."

Rafferty stopped with his foot in the air over the top step. "I beg your pardon?" he said. "An airplane came down? And it didn't have wings?"

"Yes," Mrs. Alsop said. "It's out there in the barn now. It belongs to some folks who bend iron with a hammer."

This, Rafferty thought, begins to smell like news again.

"Oh, a helicopter," he said.

Mrs. Alsop shook her head. "No, I don't think it is. It doesn't have any of those fans. But you can go out to the barn and have a look. Take him out, Alfred. Tell him to keep on the walk because it's muddy."

"Come along," Mr. Alsop said brightly. "I'd like to look the contraption over again myself."

Rafferty followed Mr. Alsop around the house on the board walk thinking he'd been mixed up with some queer people in his work, some crackpots and some screwballs, some imbeciles and some lunatics, but for sheer dumbness, these Alsops had them all beat.

"Got a lot of chickens this year," Mr. Alsop said. "All fine stock. Minorcas. Sent away for roosters and I've built a fine flock. But do you think chickens'll do very well up on a star, Mr. Rafferty?"

Rafferty involuntarily looked up at the sky and stepped off the boards into the mud.

"Up on a what?"

"I said up on a star." Mr. Alsop had reached the barn door and was trying to shove it open. "Sticks," he said. Rafferty put his shoulder to it and the door slid. When it was open a foot, Rafferty looked inside and he knew he had a story.

The object inside looked like a giant plastic balloon only half inflated so that it was globular on top and its flat bottom rested on the straw-covered floor. It was just small enough to go through the barn door. Obviously it was somebody's crackpot idea of a space ship, Rafferty thought. The headline that

The Star Ducks

flashed across his mind in thirty-six point Bodoni was "Local Farmer Builds Rocket Ship For Moon Voyage."

"Mr. Alsop," Rafferty said hopefully, "you didn't build this thing, did you?"

Mr. Alsop laughed. "Oh, no, I didn't build it. I wouldn't know how to build one of those things. Some friends of ours came in it. Gosh, I wouldn't even know how to fly one."

Rafferty looked at Mr. Alsop narrowly and he saw the man's face was serious.

"Just who are these friends of yours, Mr. Alsop?" Rafferty asked cautiously.

"Well, it sounds funny," Mr. Alsop said, "but I don't rightly know. They don't talk so very good. They don't talk at all. All we can get out of them is that their name is something about bending iron with a hammer."

Rafferty had been circling the contraption, gradually drawing closer to it. He suddenly collided with something he couldn't see. He said "ouch" and rubbed his shin.

"Oh, I forgot to tell you, Mr. Rafferty," Mr. Alsop said, "they got a gadget on it that won't let you get near, some kind of a wall you can't see. That's to keep boys away from it."

"These friends of yours, Mr. Alsop, where are they now?"

"Oh, they're over at the house," Mr. Alsop said. "You can see them if you want to. But I think you'll find it pretty hard talking to them."

"Russians?" Rafferty asked.

"Oh, no, I don't think so. They don't wear cossacks."

"Let's go," Rafferty said in a low voice and led the way across the muddy barnyard toward the house.

"These folks come here the first time about six years ago," Mr. Alsop said. "Wanted some eggs. Thought maybe they could raise chickens up where they are. Took 'em three years to get home. Eggs spoiled. So the folks turned right around and came back. This time I fixed 'em up a little brooder so they can raise chickens on the way home." He suddenly laughed. "I can just see that little contraption way out there in the sky full of chickens."

Rafferty climbed up on the back porch ahead of Mr. Alsop and went through the back door into the kitchen. Mr. Alsop stopped him before they went into the living room.

"Now, Mr. Rafferty, my wife can talk to these people better than I can, so anything you want to know you better ask her. Her and the lady get along pretty good."

"Okay," Rafferty said. He pushed Mr. Alsop gently through

the door into the living room, thinking he would play along, act naïve.

Mrs. Alsop sat in an armchair close to a circulating heater. Rafferty saw the visitors sitting side by side on the davenport, he saw them waving their long, flexible antennae delicately, he saw their lavender faces as expressionless as glass, the round eyes that seemed to be painted on.

Rafferty clutched the door facings and stared.

Mrs. Alsop turned toward him brightly.

"Mr. Rafferty," she said, "these are the people that came to see us in that airplane." Mrs. Alsop raised her finger and both the strangers bent their antennae down in her direction.

"This is Mr. Rafferty," Mrs. Alsop said. "He's a newspaper reporter. He wanted to see your airplane."

Rafferty managed to nod and the strangers curled up their antennae and nodded politely. The woman scratched her side with her left claw.

Something inside Rafferty's head was saying, you're a smart boy, Rafferty, you're too smart to be taken in. Somebody's pulling a whopping, skillful publicity scheme, somebody's got you down for a sucker. Either that or you're crazy or drunk or dreaming.

Rafferty tried to keep his voice casual.

"What did you say their names are, Mrs. Alsop?"

"Well, we don't know," Mrs. Alsop said. "You see they can only make pictures for you. They point those funny squiggly horns at you and they just think. That makes you think, too— the same thing they're thinking. I asked them what their name is and then I let them think for me. All I saw was a picture of the man hammering some iron on an anvil. So I guess their name is something like Man-Who-Bends-Iron. Maybe it's kind of like an Indian name."

Rafferty looked slyly at the people who bent iron and at Mrs. Alsop.

"Do you suppose," he said innocently, "they would talk to me—or *think* to me?"

Mrs. Alsop looked troubled.

"They'd be glad to, Mr. Rafferty. The only thing is, it's pretty hard at first. Hard for you, that is."

"I'll try it," Rafferty said. He took out a cigarette and lighted it. He held the match until it burned his fingers.

"Just throw it in the coal bucket," Mr. Alsop said.

Rafferty threw the match in the coal bucket.

"Ask these things . . . a . . . people where they come from," he said.

The Star Ducks

Mrs. Alsop smiled. "That's a very hard question. I asked them that before but I didn't get much of a picture. But I'll ask them again."

Mrs. Alsop raised her finger and both horns bent toward her and aimed directly at her head.

"This young man," Mrs. Alsop said in a loud voice like she was talking to someone hard of hearing, "wants to know where you come from."

Mr. Alsop nudged Rafferty. "Just hold up your fingers when you want your answer."

Rafferty felt like a complete idiot but he held up his finger. The woman whose husband bends iron bent her antenna down until it focused on Rafferty between the eyes. He involuntarily braced himself against the door facings. Suddenly his brain felt as though it were made of rubber and somebody was wringing and twisting and pounding it all out of shape and moulding it back together again into something new. The terror of it blinded him. He was flying through space, through a great white void. Stars and meteors whizzed by and a great star, dazzling with brilliance, white and sparkling stood there in his mind and then it went out. Rafferty's mind was released but he found himself trembling, clutching the door facings. His burning cigarette was on the floor. Mr. Alsop stooped and picked it up.

"Here's your cigarette, Mr. Rafferty. Did you get your answer?"

Rafferty was white.

"Mr. Alsop!" he said. "Mrs. Alsop! This is on the level. These creatures are really from out there in space somewhere!"

Mr. Alsop said: "Sure, they come a long way."

"Do you know what this means?" Rafferty heard his voice becoming hysterical and he tried to keep it calm. "Do you know this is the most important thing that has ever happened in the history of the world? Do you know this is . . . yes it is, it's the biggest story in the world and it's happening to me, do you understand?" Rafferty was yelling. "Where's your phone?"

"We don't have a telephone," Mr. Alsop said. "There's one down at the filling station. But these people are going to go in a few minutes. Why don't you wait and see them off? Already got their eggs and the brooder and feed on board."

"No!" Rafferty gasped. "They can't go in a few minutes! Listen. I've got to phone—I've got to get a photographer!"

Mrs. Alsop smiled.

"Well, Mr. Rafferty, we tried to get them to stay over for

supper but they have to go at a certain time. They have to catch the tide or something like that."

"It's the moon," Mr. Alsop said with authority. "It's something about the moon being in the right place."

The people from space sat there demurely, their claws folded in their laps, their antennae neatly curled to show they weren't eavesdropping on other people's minds.

Rafferty looked frantically around the room for a telephone he knew wasn't there. Got to get Joe Pegley at the city desk, Rafferty thought, Joe'll know what to do. No, no. Joe would say you're drunk.

But this is the biggest story in the world, Rafferty's brain kept saying. It's the biggest story in the world and you just stand here.

"Listen, Alsop!" Rafferty yelled. "You got a camera? Any kind of a camera. I *got* to have a camera!"

"Oh, sure," Mr. Alsop said. "I got a fine camera. It's a box camera but it takes good pictures. I'll show you some I took of my chickens."

"No, no! I don't want to see your pictures. I want the camera!"

Mr. Alsop went into the parlor and Rafferty could see him fumbling around on top of the organ.

"Mrs. Alsop!" Rafferty shouted. "I've got to ask lots of questions!"

"Ask away," Mrs. Alsop said cheerily. "They don't mind."

But what could you ask people from space? You got their names. You got what they were here for: eggs. You got where they were from. . . .

Mr. Alsop's voice came from the parlor.

"Ethel, you seen my camera?"

Mrs. Alsop sighed. "No I haven't. You put it away."

"Only trouble is," Mr. Alsop said, "haven't got any films for it."

Suddenly the people from space turned their antennae toward each other for a second and apparently coming to a mutual agreement, got up and darted here and there about the room as quick as fireflies, so fast Rafferty could scarcely see them. They scuttered out the door and off toward the barn. All Rafferty could think was: "My God, they're part bug!"

Rafferty rushed out the door, on toward the barn through the mud, screaming at the creatures to stop. But before he was half way there the gleaming plastic contraption slid out of the barn and there was a slight hiss. The thing disappeared into the low hanging clouds.

The Star Ducks

All there was left for Rafferty to see was a steaming place in the mud and a little circle of burnt earth. Rafferty sat down in the mud, a hollow, empty feeling in his middle, with the knowledge that the greatest story in the world had gone off into the sky. No pictures, no evidence, no story. He dully went over in his mind the information he had:

"Mr. and Mrs. Man-Who-Bends-Iron. . . ." It slowly dawned on Rafferty what that meant. Smith! Man-Who-Bends-Iron on an anvil. Of course that was Smith. . . . "Mr. and Mrs. Smith visited at the Alfred Alsop place Sunday. They returned home in the system of Alpha Centauri with two crates of hatching eggs."

Rafferty got to his feet and shook his head. He stood still in the mud and suddenly his eyes narrowed and you knew that Rafferty brain was working—that Rafferty brain that always came up with the story. He bolted for the house and burst in the back door.

"Alsop!" he yelled. "Did those people pay for those eggs?"

Mr. Alsop was standing on a chair in front of the china closet, still hunting for the camera.

"Oh, sure," he said. "In a way they did."

"Let me see the money!" Rafferty demanded.

"Oh, not in money," Mr. Alsop said. "They don't have any money. But when they were here six years ago they brought us some eggs of their own in trade."

"Six years ago!" Rafferty moaned. Then he started. "Eggs! What kind of eggs?"

Mr. Alsop chuckled a little. "Oh, I don't know," he said. "We called them star ducks. The eggs were star shaped. And you know we set them under a hen and the star points bothered the old hen something awful."

Mr. Alsop climbed down from the chair.

"Star ducks aren't much good though. They look something like a little hippopotamus and something like a swallow. But they got six legs. Only two of them lived and we ate them for Thanksgiving."

Rafferty's brain still worked, grasping for that single fragment of evidence that would make his city editor believe—that would make the world believe.

Rafferty leaned closer. "Mr. Alsop," he almost whispered, "you wouldn't know where the skeletons of the star ducks are?"

Mr. Alsop looked puzzled. "You mean the bones? We gave the bones to the dog. That was five years ago. Even the dog's dead now. I know where his bones are though."

Rafferty picked up his hat like a man in a daze.

"Thanks, Mr. Alsop," he said dully. "Thanks."

Rafferty stood on the porch and put on his hat. He pushed it back on his head. He stared up into the overcast; he stared until he felt dizzy like he was spiralling off into the mist, spiralling off the earth like a celestial barber pole.

Mr. Alsop opened the door and came out, wiping the dust off a box camera with his sleeve.

"Oh, Mr. Rafferty," he said. "I found the camera."

TO SERVE MAN

By Damon Knight

THE KANAMIT were not very pretty, it's true. They looked something like pigs and something like people, and that is not an attractive combination. Seeing them for the first time shocked you; that was their handicap. When a thing with the countenance of a fiend comes from the stars and offers a gift, you are disinclined to accept.

I don't know what we expected interstellar visitors to look like—those who thought about it at all, that is. Angels, perhaps, something too alien to be really awful. Maybe that's why we were all so horrified and repelled when they landed in their great ships and we saw what they really were like.

The Kanamit were short and very hairy—thick, bristly brown-gray hair all over their abominably plump bodies. Their noses were snoutlike and their eyes small, and they had thick hands of three fingers each. They wore green leather harness and green shorts, but I think the shorts were a concession to our notions of public decency. The garments were quite modishly cut, with slash pockets and half-belts in the back. The Kanamit had a sense of humor, anyhow; their clothes proved it.

There were three of them at this session of the U. N., and I can't tell you how queer it looked to see them there in the middle of a solemn Plenary Session—three fat piglike creatures in green harness and shorts, sitting at the long table below the podium, surrounded by the packed arcs of delegates from every nation. They sat correctly upright, politely watching each speaker. Their flat ears drooped over the earphones. Later on, I believe, they learned every human language, but at this time they knew only French and English.

They seemed perfectly at ease—and that, along with their humor, was a thing that tended to make me like them. I was in

---

Copyright 1950 by Damon Knight, and originally published in *Galaxy Science Fiction,* November, 1950.

the minority; I didn't think they were trying to put anything over. They said quite simply that they wanted to help us and I believed it. As a U. N. translator, of course, my opinion didn't matter, but I thought they were the best thing that ever happened to Earth.

The delegate from Argentina got up and said that his government was interested by the demonstration of a new cheap power source, which the Kanamit had made at the previous session, but that the Argentine government could not commit itself as to its future policy without a much more thorough examination.

It was what all the delegates were saying, but I had to pay particular attention to Señor Valdes, because he tended to sputter and his diction was bad. I got through the translation all right, with only one or two momentary hesitations, and then switched to the Polish-English line to hear how Gregori was doing with Janciewicz. Janciewicz was the cross Gregori had to bear, just as Valdes was mine.

Janciewicz repeated the previous remarks with a few ideological variations, and then the Secretary-General recognized the delegate from France, who introduced Dr. Denis Leveque, the criminologist, and a great deal of complicated equipment was wheeled in.

Dr. Leveque remarked that the question in many people's minds had been aptly expressed by the delegate from the U. S. S. R. at the preceding session, when he demanded, "What is the motive of the Kanamit? What is their purpose in offering us these unprecedented gifts, while asking nothing in return?"

The doctor then said, "At the request of several delegates and with the full consent of our guests, the Kanamit, my associates and I have made a series of tests upon the Kanamit with the equipment which you see before you. These tests will now be repeated."

A murmur ran through the chamber. There was a fusillade of flashbulbs, and one of the TV cameras moved up to focus on the instrument board of the doctor's equipment. At the same time, the huge television screen behind the podium lighted up, and we saw the blank faces of two dials, each with its pointer resting at zero, and a strip of paper tape with a stylus point resting against it.

The doctor's assistants were fastening wires to the temples of one of the Kanamit, wrapping a canvas-covered rubber tube around his forearm, and taping something to the palm of his right hand.

In the screen, we saw the paper tape begin to move while the stylus traced a slow zigzag pattern along it. One of the needles began to jump rhythmically; the other flipped over and stayed there, wavering slightly.

"These are the standard instruments for testing the truth of a statement," said Dr. Leveque. "Our first object, since the physiology of the Kanamit is unknown to us, was to determine whether or not they react to these tests as human beings do. We will now repeat one of the many experiments which was made in the endeavor to discover this."

He pointed to the first dial. "This instrument registers the subject's heart-beat. This shows the electrical conductivity of the skin in the palm of his hand, a measure of perspiration, which increases under stress. And this—" pointing to the tape-and-stylus device— "shows the pattern and intensity of the electrical waves emanating from his brain. It has been shown, with human subjects, that all these readings vary markedly depending upon whether the subject is speaking the truth."

He picked up two large pieces of cardboard, one red and one black. The red one was a square about a meter on a side; the black was a rectangle a meter and a half long. He addressed himself to the Kanama:

"Which of these is longer than the other?"

"The red," said the Kanama.

Both needles leaped wildly, and so did the line on the unrolling tape.

"I shall repeat the question," said the doctor. "Which of these is longer than the other?"

"The black," said the creature.

This time the instruments continued in their normal rhythm.

"How did you come to this planet?" asked the doctor.

"Walked," replied the Kanama.

Again the instruments responded, and there was a subdued ripple of laughter in the chamber.

"Once more," said the doctor, "how did you come to this planet?"

"In a spaceship," said the Kanama, and the instruments did not jump.

The doctor again faced the delegates. "Many such experiments were made," he said, "and my colleagues and myself are satisfied that the mechanisms are effective. Now," he turned to the Kanama, "I shall ask our distinguished guest to reply to the question put at the last-session by the delegate of the U. S. S. R., namely, what is the motive of the Kanamit people in offering these great gifts to the people of Earth?"

The Kanama rose. Speaking this time in English, he said,

"On my planet there is a saying, 'There are more riddles in a stone than in a philosopher's head.' The motives of intelligent beings, though they may at times appear obscure, are simple things compared to the complex workings of the natural universe. Therefore I hope that the people of Earth will understand, and believe, when I tell you that our mission upon your planet is simply this—to bring to you the peace and plenty which we ourselves enjoy, and which we have in the past brought to other races throughout the galaxy. When your world has no more hunger, no more war, no more needless suffering, that will be our reward."

And the needles had not jumped once.

The delegate from the Ukraine jumped to his feet, asking to be recognized, but the time was up and the Secretary-General closed the session.

I met Gregori as we were leaving the U. N. chamber. His face was red with excitement. "Who promoted that circus?" he demanded.

"The tests looked genuine to me," I told him.

"A circus!" he said vehemently. "A second-rate farce! If they were genuine, Peter, why was debate stifled?"

"There'll be time for debate tomorrow surely."

"Tomorrow the doctor and his instruments will be back in Paris. Plenty of things can happen before tomorrow. In the name of sanity, man, how can anybody trust a thing that looks as if it ate the baby?"

I was a little annoyed. I said, "Are you sure you're not more worried about their politics than their appearance?"

He said, "Bah," and went away.

The next day reports began to come in from government laboratories all over the world where the Kanamit's power source was being tested. They were wildly enthusiastic. I don't understand such things myself, but it seemed that those little metal boxes would give more electrical power than an atomic pile, for next to nothing and nearly forever. And it was said that they were so cheap to manufacture that everybody in the world could have one of his own. In the early afternoon there were reports that seventeen countries had already begun to set up factories to turn them out.

The next day the Kanamit turned up with plans and specimens of a gadget that would increase the fertility of any arable land by sixty to one hundred per cent. It speeded the formation of nitrates in the soil, or something. There was nothing in the headlines but the Kanamit any more. The day after that, they dropped their bombshell.

"You now have potentially unlimited power and increased food supply," said one of them. He pointed with his three-fingered hand to an instrument that stood on the table before him. It was a box on a tripod, with a parabolic reflector on the front of it. "We offer you today a third gift which is at least as important as the first two."

He beckoned to the TV men to roll their cameras into close-up position. Then he picked up a large sheet of cardboard covered with drawings and English lettering. We saw it on the large screen above the podium; it was clearly legible.

"We are informed that this broadcast is being relayed throughout your world," said the Kanama. "I wish that everyone who has equipment for taking photographs from television screens would use it now."

The Secretary-General leaned forward and asked a question sharply, but the Kanama ignored him.

"This device," he said, "projects a field in which no explosive, of whatever nature, can detonate."

There was an uncomprehending silence.

The Kanama said, "It cannot now be suppressed. If one nation has it, all must have it." When nobody seemed to understand, he explained bluntly, "There will be no more war."

That was the biggest news of the millennium, and it was perfectly true. It turned out that the explosions the Kanama was talking about included gasoline and Diesel explosions. They had simply made it impossible for anybody to mount or equip a modern army.

We could have gone back to bows and arrows, of course, but that wouldn't have satisfied the military. Not after having atomic bombs and all the rest. Besides, there wouldn't be any reason to make war. Every nation would soon have everything.

Nobody ever gave another thought to those lie-detector experiments, or asked the Kanamit what their politics were. Gregori was put out; he had nothing to prove his suspicions.

I quit my job with the U. N. a few months later, because I foresaw that it was going to die under me anyhow. U. N. business was booming at the time, but after a year or so there was going to be nothing for it to do. Every nation on Earth was well on the way to being completely self-supporting; they weren't going to need much arbitration.

I accepted a position as translator with the Kanamit Embassy, and it was there that I ran into Gregori again. I was glad to see him, but I couldn't imagine what he was doing there.

"I thought you were on the opposition," I said. "Don't tell me you're convinced the Kanamit are all right."

He looked rather shamefaced. "They're not what they look, anyhow," he said.

It was as much of a concession as he could decently make, and I invited him down to the embassy lounge for a drink. It was an intimate kind of place, and he grew confidential over the second daiquiri.

"They fascinate me," he said. "I hate them instinctively on sight, still—that hasn't changed, but I can evaluate it. You were right, obviously; they mean us nothing but good. But do you know—" he leaned across the table—"the question of the Soviet delegate was never answered."

I am afraid I snorted.

"No, really," he said. "They told us what they wanted to do —'to bring to you the peace and plenty which we ourselves enjoy.' But they didn't say *why*."

"Why do missionaries—"

"Hogwash!" he said angrily. "Missionaries have a religious motive. If these creatures do own a religion, they haven't once mentioned it. What's more, they didn't send a missionary group, they sent a diplomatic delegation—a group representing the will and policy of their whole people. Now just what have the Kanamit, as a people or a nation, got to gain from our welfare?"

I said, "Cultural—"

"Cultural cabbage-soup! No, it's something less obvious than that, something obscure that belongs to their psychology and not to ours. But trust me, Peter, there is no such thing as a completely disinterested altruism. In one way or another, they have something to gain."

"And that's why you're here," I said, "to try to find out what it is?"

"Correct. I wanted to get on one of the ten-year exchange groups to their home planet, but I couldn't; the quota was filled a week after they made the announcement. This is the next best thing. I'm studying their language, and you know that language reflects the basic assumptions of the people who use it. I've got a fair command of the spoken lingo already. It's not hard, really—some of the idioms are almost the same as the equivalents in English. And there are hints in it. I'm sure I'll get the answer eventually."

"More power," I said, and we went back to work.

I saw Gregori frequently from then on, and he kept me posted about his progress. He was highly excited about a month after that first meeting; said he'd got hold of a book of the Kanamit's and was trying to puzzle it out. They wrote in ideographs, worse than Chinese, but he was determined to fathom it if it took him years. He wanted my help.

Well, I was interested in spite of myself, for I knew it would be a long job. We spent some evenings together, working with material from Kanamit bulletin-boards and so forth, and the extremely limited English-Kanamit dictionary they issued the staff. My conscience bothered me about the stolen book, but gradually I became absorbed by the problem. Languages are my field, after all. I couldn't help being fascinated.

We got the title worked out in a few weeks. It was "How to Serve Man," evidently a handbook they were giving out to new Kanamit members of the embassy staff. They had new ones in, all the time now, a shipload about once a month; they were opening all kinds of research laboratories, clinics and so on. If there was anybody on Earth besides Gregori who still distrusted those people, he must have been somewhere in the middle of Tibet.

It was astonishing to see the changes that had been wrought in less than a year. There were no more standing armies, no more shortages, no unemployment. When you picked up a newspaper you didn't see "H-BOMB" or "V-2" leaping out at you; the news was always good. It was a hard thing to get used to. The Kanamit were working on human biochemistry, and it was known around the embassy that they were nearly ready to announce methods of making our race taller and stronger and healthier—practically a race of supermen—and they had a potential cure for heart disease and cancer.

I didn't see Gregori for a fortnight after we finished working out the title of the book; I was on a long-overdue vacation in Canada. When I got back, I was shocked by the change in his appearance.

"What on Earth is wrong, Gregori?" I asked. "You look like the very devil."

"Come down to the lounge."

I went with him, and he gulped a stiff Scotch as if he needed it.

"Come on, man, what's the matter?" I urged.

"The Kanamit have put me on the passenger list for the next exchange ship," he said. "You, too, otherwise I wouldn't be talking to you."

"Well," I said, "but—"

"They're not altruists."

"What do you mean?"

"What I told you," he said. "They're not altruists."

I tried to reason with him. I pointed out they'd made Earth

a paradise compared to what it was before. He only shook his head.

Then I said, "Well, what about those lie-detector tests?"

"A farce," he replied, without heat. "I said so at the time, you fool. They told the truth, though, as far as it went."

"And the book?" I demanded, annoyed. "What about that—'How to Serve Man'? That wasn't put there for you to read. They *mean* it. How do you explain that?"

"I've read the first page of that book," he said. "Why do you suppose I haven't slept for a week?"

I said, "Well?" and he smiled that curious, twisted smile, as if he really wanted to cry instead.

"It's a cookbook," he said.

THE FOX IN THE FOREST

By Ray Bradbury

THERE were fireworks the very first night, things that you should be afraid of perhaps, for they might remind you of other more horrible things, but these were beautiful, rockets that ascended into the ancient soft air of Mexico and shook the stars apart in blue and white fragments. Everything was good and sweet, the air was that blend of the dead and the living, of the rains and the dusts, of the incense from the church, and the brass smell of the tubas on the band stand which pulsed out vast rhythms of *La Paloma*. The church doors were thrown wide and it seemed as if a giant yellow constellation had fallen from the October sky and lay breathing fire upon the church walls; a million candles sent their color and fumes about. Newer and better fireworks scurried like tight-rope walking comets across the cool-tiled square, banged against adobe café walls, then rushed on hot wires to bash the high church tower, in which boys' naked feet alone could be seen kicking and re-kicking, clanging and tilting and re-tilting the monster bells into monstrous music. A flaming bull blundered about the plaza chasing laughing men and screaming children.

"The year is 1938," said William Travis, standing by his wife on the edge of the yelling crowd, smiling. "A good year."

The bull rushed upon them. Ducking, the couple ran, the fire-balls pelting them, past the music and riot, the church, the band, under the stars, clutching each other, laughing. The bull passed, carried lightly on the shoulders of a charging Mexican, a framework of bamboo and sulphurous gunpowder.

"I've never enjoyed myself so much in my life." Susan Travis had stopped for her breath.

"It's amazing," said William.

"It will go on, won't it?"

"All night."

"No, I mean our trip."

Copyright 1950 by Ray Bradbury, and originally published in *Collier's*, May 13, 1950.

He frowned and patted his breast-pocket. "I've enough traveler's checks for a life-time. Enjoy yourself. Forget it. They'll never find us."

"Never?"

"Never."

Now someone was setting off giant crackers, hurling them from the great bell-tolling tower of the church in a sputter of smoke, while the crowd below fell back under the threat and the crackers exploded in wonderful concussions among their dancing feet and flailing bodies. A wondrous smell of frying tortillas hung all about and in the cafés men sat at tables looking out, mugs of beer in their brown hands.

The bull was dead. The fire was out of the bamboo tubes and he was expended. The laborer lifted the framework from his shoulders. Little boys clustered to touch the magnificent papier-mâché head, the real horns.

"Let's examine the bull," said William.

As they walked past the café entrance, Susan saw the man looking out at them, a white man in a salt white suit, with a blue tie and blue shirt, and a thin, sunburnt face. His hair was blonde and straight and his eyes were blue and he watched them as they walked.

She would never have noticed him if it had not been for the bottles at his immaculate elbow; a fat bottle of crème de menthe, a clear bottle of vermouth, a flagon of cognac, and seven other bottles of assorted liqueurs, and, at his fingertips, ten small half-filled glasses from which, without taking his eyes off the street, he sipped, occasionally squinting, pressing his thin mouth shut upon the savor. In his free hand a thin Havana cigar smoked, and on a chair stoood twenty cartons of Turkish cigarettes, six boxes of cigars and some packaged colognes.

"Bill—" whispered Susan.

"Take it easy," William said. "That man's nobody."

"I saw him in the plaza this morning."

"Don't look back, keep walking, examine the papier-mâché bull—here, that's it, ask questions."

"Do you think he's from the Searchers?"

"They *couldn't* follow us!"

"They might!"

"What a nice bull," said William to the man who owned it. "He couldn't have followed us back through two hundred years, could he?"

"Watch yourself!" said William.

She swayed. He crushed her elbow tightly, steering her away.

The Fox in the Forest

"Don't faint." He smiled, to make it look good. "You'll be all right. Let's go right in that café, drink in front of him, so if he *is* what we think he is, he won't suspect."

"No, I couldn't."

"We've *got* to—come on now. And so I said to David, that's *ridiculous!*" He spoke this last in a loud voice as they went up the cafe steps.

We are here, thought Susan. Who are we? Where are we going? What do we fear? Start at the beginning, she told herself, holding to her sanity, as she felt the adobe floor underfoot.

My name is Ann Kristen, my husband's name is Roger, we were born in the year 2155 A.D. And he lived in a world that was evil. A world that was like a great ship pulling away from the shore of sanity and civilization, roaring its black horn in the night, taking two billion people with it, whether they wanted to go or not, to death, to fall over the edge of the earth and the sea into radioactive flame and madness.

They walked into the café. The man was staring at them. A phone rang.

The phone startled Susan. She remembered a phone ringing two hundred years in the future, on that blue April morning in 2155, and herself answering it:

"Ann, this is René! Have you heard? I mean about Travel In Time, Incorporated? Trips to Rome in 21 B.C., trips to Napoleon's Waterloo, any time, any place!"

"Rene, you're joking."

"No. Clinton Smith left this morning for Philadelphia in 1776. Travel In Time, Inc., arranges everything. Costs money. But *think*, to actually *see* the burning of Rome, to see Kublai Khan, Moses and the Red Sea! You've probably got an ad in your tube-mail now."

She had opened the suction mail-tube and there was the metal foil advertisement:

ROME AND THE BORGIAS!

THE WRIGHT BROTHERS AT KITTY HAWK!

Travel In Time, Inc., can costume you, put you in a crowd during the assassination of Lincoln or Caesar! We guarantee to teach you any language you need to move freely in any civilization, in any year, without friction. Latin, Greek, ancient American colloquial. Take your vacation in *Time* as well as Place!

René's voice was buzzing on the phone. "Tom and I leave

for 1492 tomorrow. They're arranging for Tom to sail with Columbus—isn't it amazing?"

"Yes," murmured Ann, stunned. "What does the government say about this Time Machine Company?"

"Oh, the police have an eye on it. Afraid people might evade the draft, run off and hide in the Past. Everyone has to leave a security bond behind, his house and belongings, to guarantee return. After all, the war's on."

"Yes, the war," murmured Ann. "The war."

Standing there, holding the phone, she had thought: Here is the chance my husband and I have talked and prayed over for so many years. We don't like this world of 2155. We want to run away from his work at the bomb factory—from my position with disease-culture units. Perhaps there is some chance for us, to escape, to run for centuries into a wild country for years where they will never find us and bring us back to burn our books, censor our thoughts, scald our minds with fear, march us, scream at us with radios . . .

The phone rang.

They were in Mexico in the year 1938.

She looked at the stained café wall.

Good workers for the Future State were allowed vacations into the Past to escape fatigue. And so she and her husband had moved back into 1938. They took a room in New York City, and enjoyed the theaters and the Statue of Liberty which still stood green in the harbor. And on the third day, they had changed their clothes and their names, and flown off to hide in Mexico.

"It *must* be him," whispered Susan, looking at the stranger seated at the table. "Those cigarettes, the cigars, the liquor. They give him away. Remember *our* first night in the Past?"

A month ago, on their first night in New York, before their flight, they had tasted all the strange drinks, bought odd foods, perfumes, cigarettes of ten dozen rare brands, for they were scarce in the Future, where war was everything. So they had made fools of themselves, rushing in and out of stores, salons, tobacconists" going up to their room to get wonderfully ill.

And now here was this stranger, doing likewise, doing a thing that only a man from the Future would do, who had been starved for liquors and cigarettes too many years.

Susan and William sat and ordered a drink.

The stranger was examining their clothes, their hair, their jewelry, the way they walked and sat.

The Fox in the Forest

"Sit easily," said William under his breath. "Look as if you've worn this clothing style all your life."

"We should never have tried to escape."

"My God," said William. "He's coming over. Let me do the talking."

The stranger bowed before them. There was the faintest tap of heels knocking together. Susan stiffened. That military sound—unmistakable as that certain ugly rap on your door at midnight.

"Mr. Kristen," said the stranger, "you did not pull up your pant legs when you sat down."

William froze. He looked at his hands lying on either leg, innocently. Susan's heart was beating swiftly.

"You've got the wrong person," said William, quickly. "My name's not Krisler."

"Kris*ten*," corrected the stranger.

"I'm William Travis," said William. "And I don't see what my pant legs have to do with you."

"Sorry." The stranger pulled up a chair. "Let us say I thought I knew you because you did *not* pull your trousers up. Everyone does. If they don't, the trousers bag quickly. I am a long way from home, Mr.—Travis—and in need of company. My name is Simms."

"Mr. Simms, we appreciate your loneliness, but we're tired. We're leaving for Acapulco tomorrow."

"A charming spot. I was just there, looking for some friends of mine. They are somewhere. I shall find them yet. Oh, is the lady a bit sick?"

"Good night, Mr. Simms."

They started out the door, William holding Susan's arm firmly. They did not look back when Mr. Simms called, "Oh, just one other thing." He paused and then slowly spoke the words:

"Twenty-one fifty-five A.D."

Susan shut her eyes and felt the earth falter under her. She kept going, into the fiery plaza, seeing nothing. . . .

They locked the door of their hotel room. And then she was crying and they were standing in the dark, and the room tilted under them. Far away, firecrackers exploded, there was laughter in the plaza.

"What a damned, loud nerve," said William. "Him sitting there, looking us up and down like animals, smoking his damn' cigarettes, drinking his drinks. I should have killed him then!" His voice was nearly hysterical. "He even had the nerve to use his real name to us. The Chief of the Searchers. And the thing about my pant legs. I should have pulled them up when I sat. It's an automatic gesture of this day and age. When I didn't do it, it set me off from the others. It made *him* think: Here's a man who never wore pants, a man used to breech-uniforms and future styles. I could kill myself for giving us away!"

"No, no, it was my walk, these high heels, that did it. Our haircuts, so new, so fresh. Everything about us odd and uneasy."

William turned on the light. "He's still testing us. He's not positive of us, not completely. We can't run out on him, then. We can't make him certain. We'll go to Acapulco, leisurely."

"Maybe he *is* sure of us, but is just playing."

"I wouldn't put it past him. He's got all the time in the world. He can dally here if he wants, and bring us back to the Future sixty seconds after we left it. He might keep us wondering for days, laughing at us."

Susan sat on the bed, wiping the tears from her face, smelling the old smell of charcoal and incense.

"They won't make a scene, will they?"

"They won't dare. They'll have to get us alone to put us in the Time Machine and send us back."

"There's a solution then," she said. "We'll never be alone, we'll always be in crowds."

Footsteps sounded outside their locked door.

They turned out the light and undressed in silence. The footsteps went away.

Susan stood by the window looking down at the plaza in the darkness. "So that building there is a church?"

"Yes."

"I've often wondered what a church looked like. It's been so long since anyone saw one. Can we visit it tomorrow?"

"Of course. Come to bed."

They lay in the dark room.

Half an hour later, their phone rang. She lifted the receiver. "Hello?"

"The rabbits may hide in the forest," said a voice, "but a fox can always find them."

She replaced the receiver and lay back straight and cold in the bed.

Outside, in the year 1938, a man played three tunes upon a guitar, one following another. . . .

During the night, she put her hand out and almost touched the year 2155. She felt her fingers slide over cool spaces of time, as over a corrugated surface, and she heard the insistent thump of marching feet, a million bands playing a million military tunes. She saw the fifty thousand rows of disease-culture in their aseptic glass tubes, her hand reaching out to them at her work in that huge factory in the future. She saw the tubes of leprosy, bubonic, typhoid, tuberculosis. She heard the great explosion and saw her hand burned to a wrinkled plum, felt it recoil from a concussion so immense that the world was lifted and let fall, and all the buildings broke and people hemorrhaged and lay silent. Great volcanoes, machines, winds, avalanches slid down to silence and she awoke, sobbing, in the bed, in Mexico, many years away . . .

In the early morning, drugged with the single hour's sleep they had finally been able to obtain, they awoke to the sound of loud automobiles in the street. Susan peered down from the iron balcony at a small crowd of eight people only now emerging, chattering, yelling, from trucks and cars with red lettering on them. A crowd of Mexicans had followed the trucks.

"*Que pasa?*" Susan called to a little boy.

The boy replied.

Susan turned back to her husband. "An American motion picture company, here on location."

"Sounds interesting." William was in the shower. "Let's watch them. I don't think we'd better leave today. We'll try to lull Simms."

For a moment, in the bright sun, she had forgotten that somewhere in the hotel, waiting, was a man smoking a thousand cigarettes, it seemed. She saw the eight loud, happy Americans below and wanted to call to them: "Save me, hide me, help me! I'm from the year 2155!"

But the words stayed in her throat. The functionaries of Travel In Time, Inc., were not foolish. In your brain, before you left on your trip, they placed a psychological block. You could tell no one your true time or birthplace, nor could you reveal any of the future to those in the past. The past and the future must be protected from each other. Only with this hindrance were people allowed to travel unguarded through the ages. The future must be protected from any change brought about by her people traveling in the past. Even if Susan wanted to with all of her heart, she could not tell any of those happy people below in the plaza who she was, or what her predicament had become.

"What about breakfast?" said William.

Breakfast was being served in the immense dining room. Ham and eggs for everyone. The place was full of tourists. The film people entered, all eight of them, six men and two women, giggling, shoving chairs about. And Susan sat near them feeling the warmth and protection they offered, even when Mr. Simms came down the lobby stairs, smoking his Turkish cigarette with great intensity. He nodded at them from a distance, and Susan nodded back, smiling, because he couldn't do anything to them here, in front of eight film people and twenty other tourists.

"Those actors," said William. "Perhaps I could hire two of them, say it was a joke, dress them in our clothes, have them drive off in our car, when Simms is in such a spot where he can't see their faces. If two people pretending to be us could lure him off for a few hours, we might make it to Mexico City. It'd take years to find us there!"

"Hey!"

A fat man, with liquor on his breath, leaned on their table.

"American tourists!" he cried. "I'm so sick of seeing Mexicans, I could kiss you!" He shook their hands. "Come on, eat with us. Misery loves company. I'm Misery, this is Miss Gloom, and Mr. and Mrs. Do-We-Hate-Mexico! We all hate it. But we're here for some preliminary shots for a damn' film. The rest of the crew arrives tomorrow. My name's Joe Melton, I'm a director, and if this ain't a hell of a country—funerals in the streets, people dying—come on, move over, join the party, cheer us up!"

Susan and William were both laughing.

"Am I funny?" Mr. Melton asked the immediate world.

"Wonderful!" Susan moved over.

Mr. Simms was glaring across the dining room at them.

She made a face at him.

Mr. Simms advanced among the tables.

"Mr. and Mrs. Travis!" he called. "I thought we were breakfasting together, alone?"

"Sorry," said William.

"Sit down, pal," said Mr. Melton. "Any friend of theirs is a pal of mine."

Mr. Simms sat. The film people talked loudly and while they talked, Mr. Simms said, quietly, "I hope you slept well."

"Did you?"

"I'm not used to spring mattresses," replied Mr. Simms, wryly. "But there are compensations. I stayed up half the night trying new cigarettes and foods. Odd, fascinating. A whole new spectrum of sensation, these ancient vices."

"We don't know what you're talking about," said Susan.

Simms laughed. "Always the play acting. It's no use. Nor is this stratagem of crowds. I'll get you alone soon enough. I'm immensely patient."

The Fox in the Forest

"Say," Mr. Melton broke in, "is this guy giving you any trouble?"

"It's all right."

"Say the word and I'll give him the bum's rush."

Melton turned back to yell at his associates. In the laughter, Mr. Simms went on: "Let us come to the point. It took me a month of tracing you through towns and cities to find you, and all of yesterday to be sure of you. If you come with me quietly, I might be able to get you off with no punishment—if you agree to go back to work on the Hydrogen-Plus bomb."

"We don't know what you're talking about."

"Stop it!" cried Simms, irritably. "Use your intelligence! You know we can't let you get away with this escape. Other people in the year 2155 might get the same idea and do the same. We need people."

"To fight your wars," said William.

"Bill!"

"It's all right, Susan. We'll talk on his terms now. We can't escape."

"Excellent," said Simms. "Really, you've both been incredibly romantic, running away from your responsibilities."

"Running away from horror."

"Nonsense. Only a war."

"What are you guys talking about?" asked Mr. Melton.

Susan wanted to tell him. But you could only speak in generalities. The psychological block in your mind allowed that. Generalities, such as Simms and William were now discussing.

"Only *the* war," said William. "Half the world dead of leprosy bombs!"

"Nevertheless," Simms pointed out, "the inhabitants of the Future resent you two hiding on a tropical isle, as it were, while they drop off the cliff into hell. Death loves death, not life. Dying people love to know that others die with them; it is a comfort to learn you are not alone in the kiln, in the grave. I am the guardian of their collective resentment against you two."

"Look at the guardian of resentments!" said Mr. Melton to his companions.

"The longer you keep me waiting, the harder it will go for you. We need you on the bomb project, Mr. Travis. Return now—no torture. Later, we'll force you to work and after you've finished the bomb, we'll try a number of complicated new devices on you, sir."

"I've got a proposition," said William. "I'll come back with you, if my wife stays here alive, safe, away from that war."

Mr. Simms debated. "All right. Meet me in the plaza in ten minutes. Pick me up in your car. Drive me to a deserted country spot. I'll have the Travel Machine pick us up there."

"Bill!" Susan held his arm tightly.

"Don't argue." He looked over at her. "It's settled." To Simms: "One thing. Last night, you could have got in our room and kidnaped us. Why didn't you?"

"Shall we say that I was enjoying myself?" replied Mr. Simms languidly, sucking his new cigar. "I hate giving up this wonderful atmosphere, this sun, this vacation. I regret leaving behind the wine and the cigarettes. Oh, how I regret it. The plaza then, in ten minutes. Your wife will be protected and may stay here as long as she wishes. Say your good-bys."

Mr. Simms arose and walked out.

"There goes Mr. Big-Talk!" yelled Mr. Melton at the departing gentleman. He turned and looked at Susan. "Hey. Someone's crying. Breakfast's no time for people to cry, now *is* it?"

At nine fifteen, Susan stood on the balcony of their room, gazing down at the plaza. Mr. Simms was seated there, his neat legs crossed, on a delicate bronze bench. Biting the tip from a cigar, he lighted it tenderly.

Susan heard the throb of a motor, and far up the street, out of a garage and down the cobbled hill, slowly, came William in his car.

The car picked up speed. Thirty, now forty, now fifty miles an hour. Chickens scattered before it.

Mr. Simms took off his white Panama hat and mopped his pink forehead, put his hat back on, and then saw the car.

It was rushing sixty miles an hour, straight on for the plaza.

"William!" screamed Susan.

The car hit the low plaza curb, thundering, jumped up, sped across the tiles toward the green bench where Mr. Simms now dropped his cigar, shrieked, flailed his hands, and was hit by the car. His body flew up and up in the air, and down, crazily, into the street.

On the far side of the plaza, one front wheel broken, the car stopped. People were running.

Susan went in and closed the balcony doors.

They came down the Official Palace steps together, arm in arm, their faces pale, at twelve noon.

"*Adiós, señor*," said the mayor behind them. "*Señora.*"

They stood in the plaza where the crowd was pointing at the blood.

The Fox in the Forest

"Will they want to see you again?" asked Susan.

"No, we went over and over it. It was an accident. I lost control of the car. I wept for them. God knows I had to get my relief out somewhere. I *felt* like weeping. I hated to kill him. I've never wanted to do anything like that in my life."

"They won't prosecute you?"

"They talked about it, but no. I talked faster. They believe me. It was an accident. It's over."

"Where will we go? Mexico City?"

"The car's in the repair shop. It'll be ready at four this afternoon. Then we'll get the hell out."

"Will we be followed? Was Simms working alone?"

"I don't know. We'll have a little head start on them, I think."

The film people were coming out of the hotel as they approached. Mr. Melton hurried up, scowling. "Hey, I heard what happened. Too bad. Everything okay now? Want to get your minds off it? We're doing some preliminary shots up the street. You want to watch, you're welcome. Come on, do you good."

They went.

They stood on the cobbled street while the film camera was being set up. Susan looked at the road leading down and away, at the highway going to Acapulco and the sea, past pyramids and ruins and little adobe towns with yellow walls, blue walls, purple walls and flaming bougainvillea. She thought: We shall take the roads, travel in clusters and crowds, in markets, in lobbies, bribe police to sleep near, keep double locks, but always the crowds, never alone again, always afraid the next person who passes might be another Simms. Never knowing if we've tricked and lost the Searchers. And always up ahead, in the Future, they'll wait for us to be brought back, waiting with their bombs to burn us and disease to rot us, and their police to tell us to roll over, turn around, jump through the hoop. And so we'll keep running through the forest, and we'll never ever stop or sleep well again in our lives.

A crowd gathered to watch the film being made. And Susan watched the crowd and the streets.

"Seen anyone suspicious?"

"No. What time is it?"

"Three o'clock. The car should be almost ready."

The test film was finished at three forty-five. They all walked down to the hotel, talking. William paused at the garage. "The car'll be ready at six," he said, coming out.

"But no later than that?"

"It'll be ready, don't worry."

In the hotel lobby they looked around for other men traveling alone, men who resembled Mr. Simms, men with new haircuts and too much cigarette smoke and cologne smell about them, but the lobby was empty. Going up the stairs, Mr. Melton said, "Well, it's been a long, hard day. Who'd like to put a header on it. Martini? Beer?"

"Maybe one."

The whole crowd pushed into Mr. Melton's room and the drinking began.

"Watch the time," said William.

Time, thought Susan, if only they had time. All she wanted was to sit in the plaza all of a long, bright day in spring, with not a worry or a thought, with the sun on her face and arms, her eyes closed, smiling at the warmth—and never move, but just sleep in the Mexican sun . . .

Mr. Melton opened the champagne.

"To a very beautiful lady, lovely enough for films," he said, toasting Susan. "I might even give you a test."

She laughed.

"I mean it," said Melton. "You're very nice. I could make you a movie star."

"And take me to Hollywood?"

"Get the hell out of Mexico, sure!"

Susan glanced at William, and he lifted an eyebrow and nodded. It would be a change of scene, clothing, locale, name perhaps, and they would be traveling with eight other people, a good shield against any interference from the future.

"It sounds wonderful," said Susan.

She was feeling the champagne now, the afternoon was slipping by, the party was whirling about her, she felt safe and good and alive and truly happy for the first time in many years.

"What kind of film would my wife be good for?" asked William, refilling his glass.

Melton appraised Susan. The party stopped laughing and listened.

"Well, I'd like to do a story of suspense," said Melton. "A story of a man and wife, like yourselves."

"Go on."

"Sort of a war story, maybe," said the director, examining the color of his drink against the sunlight.

Susan and William waited.

"A story about a man and wife who live in a little house on a little street in the year 2155, maybe," said Melton. "This is ad lib, understand. But this man and wife are faced with a terrible war, Super-Plus Hydrogen bombs, censorship, death, in that year and—here's the gimmick—they escape into the past, followed by a man who they think is evil, but who is only trying to show them what their Duty is."

William dropped his glass to the floor.

Mr. Melton continued. "And this couple take refuge with a group of film people whom they learn to trust. Safety in numbers, they say to themselves."

Susan felt herself slip down into a chair. Everyone was watching the director. He took a little sip of wine. "Ah, that's a fine wine. Well, this man and woman, it seems, don't realize how important they are to the future. The man, especially, is the keystone to a new bomb metal. So the Searchers, let's call them, spare no trouble or expense to find, capture and take home the man and wife, once they get them totally alone, in a hotel room, where no one can see. Strategy. The Searchers work alone, or in groups of eight. One trick or another will do it. Don't you think it would make a wonderful film, Susan? Don't you, Bill?" He finished his drink.

Susan sat with her eyes straight ahead.

"Have a drink?" said Mr. Melton.

William's gun was out and fired three times, and one of the men fell, and the others ran forward. Susan screamed. A hand was clamped to her mouth. Now the gun was on the floor and William was struggling with the men holding him.

Mr. Melton said, "Please," standing there where he had stood, blood showing on his fingers. "Let's not make matters worse."

Someone pounded on the hall door.

"Let me in!"

"The manager," said Mr. Melton, dryly. He jerked his head. "Everyone, let's move!"

"Let me in. I'll call the police!"

Susan and William looked at each other quickly, and then at the door.

"The manager wishes to come in," said Mr. Melton. "Quick!"

A camera was carried forward. From it shot a blue light which encompassed the room instantly. It widened out and the people of the party vanished, one by one.

"Quickly!"

Outside the window in the instant before she vanished, Susan saw the green land and the purple and yellow and blue and crimson walls and the cobbles flowing like a river, a man upon a burro riding into the warm hills, a boy drinking orange pop,

She could feel the sweet liquid in her throat; she could see a man standing under a cool plaza tree with a guitar, could feel her hand upon the strings. And, far away, she could see the sea, the blue and tender sea; she could feel it roll her over and take her in.

And then she was gone. Her husband was gone.

The door burst wide. The manager and his staff rushed in. The room was empty.

"But they were just here! I saw them come in, and now—gone!" cried the manager. "The windows are covered with iron grating; they couldn't get out that way!" . . .

In the late afternoon, the priest was summoned and they opened the room again and aired it out, and had him sprinkle holy water through each corner and give it his cleansing.

"What shall we do with these?" asked the charwoman.

She pointed to the closet, where there were sixty-seven bottles of chartreuse, cognac, *crème de cacao*, absinthe, vermouth, tequila, 106 cartons of Turkish cigarettes, and 198 yellow boxes of fifty-cent pure Havana-filler cigars. . . .

NINE-FINGER JACK

By Anthony Boucher

JOHN SMITH is an unexciting name to possess, and there was of course no way for him to know until the end of his career that he would be forever famous among connoisseurs of murder as Nine-finger Jack. But he did not mind the drabness of Smith; he felt that what was good enough for the great George Joseph was good enough for him.

Not only did John Smith happily share his surname with George Joseph; he was proud to follow the celebrated G. J. in profession and even in method. For an attractive and plausible man of a certain age, there are few more satisfactory sources of income than frequent and systematic widowerhood; and of all the practitioners who have acted upon this practical principle, none have improved upon George Joseph Smith's sensible and unpatented Brides-in-the-Bath method.

John Smith's marriage to his ninth bride, Hester Pringle, took place on the morning of May the thirty-first. On the evening of May the thirty-first John Smith, having spent much of the afternoon pointing out to friends how much the wedding had excited Hester and how much he feared the effect on her notoriously weak heart, entered the bathroom and, with the careless ease of the practiced professional, employed five of his fingers to seize Hester's ankles and jerk her legs out of the tub while with the other five fingers he gently pressed her face just below water level.

So far all had proceeded in the conventional manner of any other wedding night; but the ensuing departure from ritual was such as to upset even John Smith's professional bathside manner. The moment Hester's face and neck were submerged below water, she opened her gills.

In his amazement, John released his grasp upon both ends of his bride. Her legs descended into the water and her face rose above it. As she passed from the element of water to that of air, her gills closed and her mouth opened.

Copyright, 1951, by Esquire, Inc., and originally published in *Esquire*, May, 1951.

"I suppose," she observed, "that in the intimacy of a long marriage you would eventually have discovered in any case that I am a Venusian. It is perhaps as well that the knowledge came early, so that we may lay a solid basis for understanding."

"Do you mean," John asked, for he was a precise man, "that you are a native of the planet Venus?"

"I do," she said. "You would be astonished to know how many of us there are already among you."

"I am sufficiently astonished," said John, "to learn of one. Would you mind convincing me that I did indeed see what I thought I saw?"

Obligingly, Hester lowered her head beneath the water. Her gills opened and her breath bubbled merrily. "The nature of our planet," she explained when she emerged, "has bred as its dominant race our species of amphibian mammals, in all other respects superficially identical with *homo sapiens.* You will find it all but impossible to recognize any of us, save perhaps by noticing those who, to avoid accidental opening of the gills, refuse to swim. Such concealment will of course be unnecessary soon when we take over complete control of your planet."

"And what do you propose to do with the race that already controls it?"

"Kill most of them, I suppose," said Hester; "and might I trouble you for that towel?"

"That," pronounced John, with any handcraftsman's abhorrence of mass production, "is monstrous. I see my duty to my race: I must reveal all."

"I am afraid," Hester observed as she dried herself, "that you will not. In the first place, no one will believe you. In the second place, I shall then be forced to present to the authorities the complete dossier which I have gathered on the cumulatively interesting deaths of your first eight wives, together with my direct evidence as to your attempt this evening."

John Smith, being a reasonable man, pressed the point no further. "In view of this attempt," he said, "I imagine you would like either a divorce or an annulment."

"Indeed I should not," said Hester. "There is no better cover for my activities than marriage to a member of the native race. In fact, should you so much as mention divorce again, I shall be forced to return to the topic of that dossier. And now, if you will hand me that robe, I intend to do a little telephoning. Some of my better-placed colleagues will need to know my new name and address."

Nine-Finger Jack

As John Smith heard her ask the long-distance operator for Washington, D. C., he realized with regretful resignation that he would be forced to depart from the methods of the immortal George Joseph.

Through the failure of the knife, John Smith learned that Venusian blood has extraordinary quick-clotting powers and Venusian organs possess an amazingly rapid system of self-regeneration. And the bullet taught him a further peculiarity of the blood: that it dissolves lead—in fact thrives upon lead.

His skill as a cook was quite sufficient to disguise any of the commoner poisons from human taste; but the Venusian palate not only detected but relished most of them. Hester was particularly taken with his tomato aspic *à l'arsénique* and insisted on his preparing it in quantity for a dinner of her friends, along with his *sole amandine* to which the prussic acid lent so distinctively intensified a flavor and aroma.

While the faintest murmur of divorce, even after a year of marriage, evoked from Hester a frowning murmur of "Dossier . . ." the attempts at murder seemed merely to amuse her; so that finally John Smith was driven to seek out Professor Gillingsworth at the State University, recognized as the ultimate authority (on this planet) on life on other planets.

The professor found the query of much theoretical interest. "From what we are able to hypothesize of the nature of Venusian organisms," he announced, "I can almost assure you of their destruction by the forced ingestion of the best Beluga caviar, in doses of no less than one-half pound per diem."

Three weeks of the suggested treatment found John Smith's bank account seriously depleted and his wife in perfect health.

"That dear Gilly!" she laughed one evening. "It was so nice of him to tell you how to kill me; it's the first time I've had enough of caviar since I came to Earth. It's so dreadfully expensive."

"You mean," John demanded, "that Professor Gillingsworth is. . . ."

She nodded.

"And all that money!" John protested. "You do not realize, Hester, how unjust you are. You have deprived me of my income and I have no other source."

"Dossier," said Hester through a mouthful of caviar.

America's greatest physiologist took an interest in John Smith's problem. "I should advise," he said, "the use of crystallized carbon placed directly in contact with the sensitive gill area."

"In other words, a diamond necklace?" John Smith asked. He seized a water carafe, hurled its contents at the physiologist's neck, and watched his gills open.

The next day John purchased a lapel flower through which water may be squirted—an article which he thenceforth found invaluable for purposes of identification.

The use of this flower proved to be a somewhat awkward method of starting a conversation and often led the conversation into unintended paths; but it did establish a certain clarity in relations.

It was after John had observed the opening of the gills of a leading criminal psychiatrist that he realized where he might find the people who could really help him.

From then on, whenever he could find time to be unobserved while Hester was engaged in her activities preparatory to world conquest, he visited insane asylums, announced that he was a free-lance feature writer, and asked if they had any inmates who believed that there were Venusians at large upon earth and planning to take it over.

In this manner he met many interesting and attractive people, all of whom wished him godspeed in his venture, but pointed out that they would hardly be where they were if all of their own plans for killing Venusians had not miscarried as hopelessly as his.

From one of these friends, who had learned more than most because his Venusian wife had made the error of falling in love with him (an error which led to her eventual removal from human society), John Smith ascertained that Venusians may indeed be harmed and even killed by many substances on their own planet, but seemingly by nothing on ours—though (his) wife had once dropped a hint that one thing alone on earth could prove fatal to the Venusian system.

At last John Smith visited an asylum whose director announced that they had an inmate who thought he *was* a Venusian.

When the director had left them, a squirt of the lapel flower verified the claimant's identity.

"I am a member of the Conciliationist Party," he explained, "the only member who has ever reached this earth. We believe that Earthmen and Venusians can live at peace as all men should, and I shall be glad to help you destroy all members of the opposition party.

"There is one substance on this earth which is deadly poison to any Venusian. Since in preparing and serving the dish best suited to its administration you must be careful to wear gloves, you should begin your campaign by wearing gloves at all meals . . ."

This mannerism Hester seemed willing to tolerate for the security afforded her by her marriage and even more particularly for the delights of John's skilled preparation of such dishes as spaghetti *all'aglio ed all'arsenico* which is so rarely to be had in the average restaurant.

Two weeks later John finally prepared the indicated dish: ox tail according to the richly imaginative recipe of Simon Templar, with a dash of deadly nightshade added to the other herbs specified by The Saint. Hester had praised the recipe, devoured two helpings, expressed some wonder as to the possibility of gills in its creator, whom she had never met, and was just nibbling at the smallest bones when, as the Conciliationist had foretold, she dropped dead.

Intent upon accomplishing his objective, John had forgotten the dossier, nor ever suspected that it was in the hands of a gilled lawyer who had instructions to pass it on in the event of Hester's death.

Even though that death was certified as natural, John rapidly found himself facing trial for murder, with seven other states vying for the privilege of the next opportunity should this trial fail to end in a conviction.

With no prospect in sight of a quiet resumption of his accustomed profession, John Smith bared his knowledge and acquired his immortal nickname. The result was a period of intense prosperity among manufacturers of squirting lapel flowers, bringing about the identification and exposure of the gilled masqueraders.

But inducing them, even by force, to ingest the substance poisonous to them was more difficult. The problem of supply and demand was an acute one, in view of the large number of the Venusians and the small proportion of members of the human race willing to perform the sacrifice made by Nine-finger Jack.

It was that great professional widower and amateur chef himself who solved the problem by proclaiming in his death cell his intention to bequeath his body to the eradication of Venusians, thereby pursuing after death the race which had ruined his career.

The noteworthy proportion of human beings who promptly followed his example in their wills has assured us of permanent protection against future invasions, since so small a quantity of the poison is necessary in each individual case; after all, one finger sufficed for Hester.

DARK INTERLUDE

By Mack Reynolds and Fredric Brown

SHERIFF BEN RAND'S *eyes were grave. He said, "Okay, boy. You feel kind of jittery; that's natural. But if your story's straight, don't worry. Don't worry about nothing. Everything'll be all right, boy."*

"It was three hours ago, Sheriff," Allenby said. "I'm sorry it took me so long to get into town and that I had to wake you up. But Sis was hysterical a while. I had to try and quiet her down, and then I had trouble starting the jalopy."

"Don't worry about waking me up, boy. Being sheriff's a full-time job. And it ain't late, anyway; I just happened to turn in early tonight. Now let me get a few things straight. You say your name's Lou Allenby. That's a good name in these parts, Allenby. You kin of Rance Allenby, used to run the feed business over in Cooperville? I went to school with Rance . . . Now about the fella who said he come from the future . . ."

The Presidor of the Historical Research Department was skeptical to the last. He argued, "I am still of the opinion that the project is not feasible. There are paradoxes involved which present insurmountable—"

Doctor Matthe, the noted physicist, interrupted politely, "Undoubtedly, sir, you are familiar with the Dichotomy?"

The presidor wasn't, so he remained silent to indicate that he wanted an explanation.

"Zeno propounded the Dichotomy. He was a Greek philosopher of roughly five hundred years before the ancient prophet whose birth was used by the primitives to mark the beginning of their calendar. The Dichotomy states that it is impossible to cover any given distance. The argument: First, half the distance must be traversed, then half of the remaining

---

Copyright, 1951, by World Editions, Inc., and originally published in *Galaxy Science Fiction*, January, 1951.

Dark Interlude

distance, then again half of what remains, and so on. It follows that some portion of the distance to be covered always remains, and therefore motion is impossible."

"Not analagous," the presidor objected. "In the first place, your Greek assumed that any totality composed of an infinite number of parts must, itself, be infinite, whereas we know that an infinite number of elements make up a finite total. Besides—"

Matthe smiled gently and held up a hand. "Please, sir, don't misunderstand me. I do not deny that today we understand Zeno's paradox. But believe me, for long centuries the best minds the human race could produce could not explain it."

The presidor said tactfully, "I fail to see your point, Doctor Matthe. Please forgive my inadequacy. What possible connection has this Dichotomy of Zeno's with your projected expedition into the past?"

"I was merely drawing a parallel, sir. Zeno conceived the paradox proving that it was impossible to cover any distance, nor were the ancients able to explain it. But did that prevent them from covering distances? Obviously not. Today, my assistants and I have devised a method to send our young friend here, Jan Obreen, into the distant past. The paradox is immediately pointed out—suppose he should kill an ancestor or otherwise change history? I do not claim to be able to explain how this apparent paradox is overcome in time travel; all I know is that time travel *is* possible. Undoubtedly, better minds than mine will one day resolve the paradox, but until then we shall continue to utilize time travel, paradox or not."

Jan Obreen had been sitting, nervously quiet, listening to his distinguished superiors. Now he cleared his throat and said, "I believe the hour has arrived for the experiment."

The presidor shrugged his continued disapproval, but dropped the conversation. He let his eyes scan doubtfully the equipment that stood in the corner of the laboratory.

Matthe shot a quick glance at the time piece, then hurried last minute instructions to his student.

"We've been all over this before, Jan, but to sum it up—You should appear approximately in the middle of the so-called Twentieth Century; exactly where, we don't know. The language will be Amer-English, which you have studied thoroughly; on that count you should have little difficulty. You will appear in the United States of North America, one of the ancient nations—as they were called—a political division of whose purpose we are not quite sure. One of the designs of your expedition will be to determine why the human race at that time split itself into scores of states, rather than having but one government.

"You will have to adapt yourself to the conditions you find, Jan. Our histories are so vague that we can help you but little in information on what to expect."

The presidor put in, "I am extremely pessimistic about this, Obreen, yet you have volunteered and I have no right to interfere. Your most important task is to leave a message that will come down to us; if you are successful, other attempts will be made to still other periods in history. If you fail—"

"He won't fail," Matthe said.

The presidor shook his head and grasped Obreen's hand in farewell.

Jan Obreen stepped to the equipment and mounted the small platform. He clutched the metal grips on the instrument panel somewhat desperately, hiding to the best of his ability the shrinking inside himself.

The sheriff said, "Well, this fella—you say he told you he came from the future?"

Lou Allenby nodded. "About four thousand years ahead. He said it was the year thirty-two hundred and something, but that it was about four thousand years from now; they'd changed the numbering system meanwhile."

"And you didn't figure it was hogwash, boy? From the way you talked, I got the idea that you kind of believed him."

The other wet his lips. "I kind of believed him," he said doggedly. "There was something about him; he was different. I don't mean physically, that he couldn't pass for being born now, but there was . . . something different. Kind of, well, like he was at peace with himself; gave the impression that where he came from everybody was. And he was smart, smart as a whip. And he wasn't crazy, either."

"And what was he doing back here, boy?" The sheriff's voice was gently caustic.

"He was—some kind of student. Seems from what he said that almost everybody in his time was a student. They'd solved all the problems of production and distribution, nobody had to worry about security; in fact they didn't seem to worry about any of the things we do now." There was a trace of wistfulness in Lou Allenby's voice. He took a deep breath and went on. "He'd come back to do research in our time. They didn't know much about it, it seems. Something had happened in between—there was a bad period of several hun-

Dark Interlude

dred years—and most books and records had been lost. They had a few, but not many. So they didn't know much about us and they wanted to fill in what they didn't know."

"You believed all that, boy? Did he have any proof?"

It was the dangerous point; this was where the prime risk lay. They had had, for all practical purposes, no knowledge of the exact contours of the land, forty centuries back, nor knowledge of the presence of trees or buildings. If he appeared at the wrong spot, it might well mean instant death.

Jan Obreen was fortunate; he didn't hit anything. It was, in fact, the other way around. He came out ten feet in the air over a plowed field. The fall was nasty enough, but the soft earth protected him; one ankle seemed sprained, but not too badly. He came painfully to his feet and looked around.

The presence of the field alone was sufficient to tell him that the Matthe process was at least partially successful. He was far before his own age. Agriculture was still a necessary component of human economy, indicating a definitely earlier civilization than his own.

Approximately half a mile away was a densely wooded area; not a park, nor even a planned forest to house the controlled wild life of his time. A haphazardly growing wooded area—almost unbelievable. But, then, he must grow used to the unbelievable; of all the historic periods, this was the least known. Much would be strange.

To his right, a few hundred yards away, was a wooden building. It was, undoubtedly, a human dwelling despite its primitive appearance. There was no use putting it off; contact with his fellow man would have to be made. He limped awkwardly toward his meeting with the Twentieth Century.

The girl had evidently not observed his precipitate arrival, but by the time he arrived in the yard of the farm house, she had come to the door to greet him.

Her dress was of another age, for in his era the clothing of the feminine portion of the race was not designed to lure the male. Hers, however, was bright and tasteful with color, and it emphasized the youthful contours of her body. Nor was it her dress alone that startled him. There was a touch of color on her lips that he suddenly realized couldn't have been achieved by nature. He had read that primitive women used colors, paints and pigments of various sorts, upon their faces—somehow or other, now that he witnessed it, he was not repelled.

She smiled, the red of her mouth stressing the even whiteness of her teeth. She said, "It would've been easier to come down the road 'stead of across the field." Her eyes took him in, and, had he been more experienced, he could have read interested approval in them.

He said, studiedly, "I am afraid that I am not familiar with your agricultural methods. I trust I have not irrevocably damaged the products of your horticultural efforts."

Susan Allenby blinked at him. "My," she said softly, a distant hint of laughter in her voice, "somebody sounds like maybe they swallowed a dictionary." Her eyes widened suddenly, as she noticed him favoring his left foot. "Why, you've hurt yourself. Now you come right on into the house and let me see if I can't do something about that. Why—"

He followed her quietly, only half hearing her words. Something—something phenomenal—was growing within Jan Obreen, affecting oddly and yet pleasantly his metabolism.

He knew now what Matthe and the presidor meant by paradox.

The sheriff said, "Well, you were away when he got to your place—however he got there?"

Lou Allenby nodded. "Yes, that was ten days ago. I was in Miami taking a couple of weeks' vacation. Sis and I each get away for a week or two every year, but we go at different times, partly because we figure it's a good idea to get away from one another once in a while anyway."

"Sure, good idea, boy. But your Sis, she believed this story of where he came from?"

"Yes. And, Sheriff, she had proof. I wish I'd seen it too. The field he landed in was fresh plowed. After she'd fixed his ankle she was curious enough, after what he'd told her, to follow his footsteps through the dirt back to where they'd started. And they ended, or, rather, started, right smack in the middle of a field, with a deep mark like he'd fallen there."

"Maybe he came from an airplane, in a parachute, boy. Did you think of that?"

"I thought of that, and so did Sis. She says that if he did he must've swallowed the parachute. She could follow his steps every bit of the way—it was only a few hundred yards—and there wasn't any place he could've hidden or buried a parachute."

The sheriff said, "They got married right away, you say?"

"Two days later. I had the car with me, so Sis hitched the team and drove them into town—he didn't know how to drive horses—and they got married."

"See the license, boy? You sure they was really—"

Lou Allenby looked at him, his lips beginning to go white, and the sheriff said hastily, "All right, boy, I didn't mean it that way. Take it easy, boy."

Susan had sent her brother a telegram telling him all about it, but he'd changed hotels and somehow the telegram hadn't been forwarded. The first he knew of the marriage was when he drove up to the farm almost a week later.

He was surprised, naturally, but John O'Brien—Susan had altered the name somewhat—seemed likable enough. Handsome, too, if a bit strange, and he and Susan seemed head over heels in love.

Of course, he didn't have any money, they didn't use it in his day, he had told them, but he was a good worker, not at all soft. There was no reason to suppose that he wouldn't make out all right.

The three of them planned, tentatively, for Susan and John to stay at the farm until John had learned the ropes somewhat. Then he expected to be able to find some manner in which to make money—he was quite optimistic about his ability in that line—and spending his time traveling taking Susan with him. Obviously, he'd be able to learn about the present that way.

The important thing, the all-embracing thing, was to plan some message to get to Doctor Matthe and the presidor. If this type of research was to continue, all depended upon him.

He explained to Susan and Lou that it was a one-way trip. That the equipment worked only in one direction, that there was travel to the past, but not to the future. He was a voluntary exile, fated to spend the rest of his life in this era. The idea was that when he'd been in this century long enough to describe it well, he'd write up his report and put it in a box he'd have especially made to last forty centuries and bury it where it could be dug up—in a spot that had been determined in the future. He had the exact place geographically.

He was quite excited when they told him about the time capsules that had been buried elsewhere. He knew that they had never been dug up and planned to make it part of his report so the men of the future could find them.

They spent their evenings in long conversations, Jan telling of his age and what he knew of all the long centuries in between. Of the long fight upward and man's conquests in the fields of science, medicine and in human relations. And they telling him of theirs, describing the institutions, the ways of life which he found so unique.

Lou hadn't been particularly happy about the precipitate marriage at first, but he found himself warming to Jan. Until . . .

The sheriff said, "And he didn't tell you what he was till this evening?"

"That's right."

"Your sister heard him say it? She'll back you up?"

"I . . . I guess she will. She's upset now, like I said, kind of hysterical. Screams that she's going to leave me and the farm. But she heard him say it, Sheriff. He must of had a strong hold on her, or she wouldn't be acting the way she is."

"Not that I doubt your word, boy, about a thing like that, but it'd be better if she heard it too. How'd it come up?"

"I got to asking him some questions about things in his time and after a while I asked him how they got along on race problems and he acted puzzled and then said he remembered something about races from history he'd studied, but that there weren't any races then.

"He said that by his time—starting after the war of something-or-other, I forget its name—all the races had blended into one. That the whites and the yellows had mostly killed one another off and that Africa had dominated the world for a while, and then all the races had begun to blend into one by colonization and intermarriage and that by his time the process was complete. I just stared at him and asked him, 'You mean you got nigger blood in you?' and he said, just like it didn't mean anything, 'At least one-fourth.'"

"Well, boy, you did just what you had to do," the sheriff told him earnestly, "no doubt about it."

"I just saw red. He'd married Sis; he was sleeping with her. I was so crazy-mad I don't even remember getting my gun."

"Well, don't worry about it, boy. You did right."

"But I feel like hell about it. He didn't know."

"Now that's a matter of opinion, boy. Maybe you swallowed a little too much of this hogwash. Coming from the future— huh! These niggers'll think up the damnedest tricks to pass themself off as white. What kind of proof for his story is that mark on the ground? Hogwash, boy. Ain't nobody coming from the future or going there neither. We can just quiet this up so it won't never be heard of nowhere. It'll be like it never happened."

GENERATION OF NOAH

By William Tenn

THAT was the day Plunkett heard his wife screaming guardedly to their youngest boy.

He let the door of the laying house slam behind him, forgetful of the nervously feeding hens. She had, he realized, cupped her hands over her mouth so that only the boy would hear.

"Saul! You, *Saul!* Come back, come right back this instant. Do you want your father to catch you out there on the road? Saul!"

The last shriek was higher and clearer, as if she had despaired of attracting the boy's attention without at the same time warning the man.

Poor Ann!

Gently, rapidly, Plunkett *shh'd* his way through the bustling and hungry hens to the side door. He came out facing the brooder run and broke into a heavy, unathletic trot.

They have the responsibility after Ann and me, Plunkett told himself. Let them watch and learn again. He heard the other children clatter out of the feed house. Good!

"Saul!" his wife's voice shrilled unhappily. "Saul, your father's coming!"

Ann came out of the front door and paused. "Elliot," she called at his back as he leaped over the flush well-cover. "Please, I don't feel well."

A difficult pregnancy, of course, and in her sixth month. But that had nothing to do with Saul. Saul knew better.

At the last frozen furrow of the truck garden Plunkett gave himself a moment to gather the necessary air for his lungs. Years ago, when Von Rundstedt's Tigers roared through the Bulge, he would have been able to dig a foxhole after such a run. Now, he was just winded. Just showed you: such a short distance from the far end of the middle

---

Copyright, 1951, by Farrell Publishing Corp., and originally published in *Suspense*, Spring, 1951.

chicken house to the far end of the vegetable garden—merely crossing four acres—and he was winded. And consider the practice he'd had.

He could just about see the boy idly lifting a stick to throw for the dog's pleasure. Saul was in the further ditch, well past the white line his father had painted across the road.

"Elliot," his wife began again. "He's only six years old. He—"

Plunkett drew his jaws apart and let breath out in a bellyful of sound. "Saul! Saul Plunkett!" he bellowed. "Start running!"

He knew his voice had carried. He clicked the button on his stopwatch and threw his right arm up, pumping his clenched fist.

The boy *had* heard the yell. He turned, and, at the sight of the moving arm that meant the stopwatch had started, he dropped the stick. But, for the fearful moment, he was too startled to move.

Eight seconds. He lifted his lids slightly. Saul had begun to run. But he hadn't picked up speed, and Rusty skipping playfully between his legs threw him off his stride.

Ann had crossed the garden laboriously and stood at his side, alternately staring over his jutting elbow at the watch and smiling hesitantly sidewise at his face. She shouldn't have come out in her thin house-dress in November. But it was good for Ann. Plunkett kept his eyes stolidly on the unemotional second hand.

One minute forty.

He could hear the dog's joyful barks coming closer, but as yet there was no echo of sneakers slapping the highway. Two minutes. He wouldn't make it.

The old bitter thoughts came crowding back to Plunkett. A father timing his six-year-old son's speed with the best watch he could afford. This, then, was the scientific way to raise children in Earth's most enlightened era. Well, it was scientific . . . in keeping with the very latest discoveries. . . .

Two and a half minutes. Rusty's barks didn't sound so very far off. Plunkett could hear the desperate pad-pad-pad of the boy's feet. He might make it at that. If only he could!

"*Hurry*, Saul," his mother breathed. "You can make it."

Plunkett looked up in time to see his son pound past, his jeans already darkened with perspiration. "Why doesn't he breathe like I told him?" he muttered. "He'll be out of breath in no time."

Halfway to the house, a furrow caught at Saul's toes. As he sprawled, Ann gasped. "You can't count that, Elliot. He tripped."

"Of course he tripped. He should count on tripping."

"Get up, Saulie," Herbie, his older brother, screamed from the garage where he stood with Louise Dawkins, one pail of eggs between them. "Get up and run! This corner here! You can make it!"

The boy weaved to his feet, and threw his body forward again. Plunkett could hear him sobbing. He reached the cellar steps—and literally plunged down.

Plunkett pressed the stopwatch and the second hand halted. Three minutes thirteen seconds.

He held the watch up for his wife to see. "Thirteen seconds, Ann."

Her face wrinkled.

He walked to the house. Saul crawled back up the steps, fragments of unrecovered breath rattling in his chest. He kept his eyes on his father.

"Come here, Saul. Come right here. Look at the watch. Now, what do you see?"

The boy stared intently at the watch. His lips began twisting; startled tears writhed down his stained face. "More— more than three m-minutes, poppa?"

"More than three minutes, Saul. Now, Saul—don't cry, son; it isn't any use—Saul, what would have happened when you got to the steps?"

A small voice, pitifully trying to cover its cracks: "The big doors would be shut."

"The big doors would be shut. You would be locked outside. Then what would have happened to you? Stop crying. Answer me!"

"Then, when the bombs fell, I'd—I'd have no place to hide. I'd burn like the head of a match. An'—an' the only thing left of me would be a dark spot on the ground, shaped like my shadow. An'—an'—"

"And the radioactive dust," his father helped with the catechism.

"Elliot—" Ann sobbed behind him. "I don't—"

"*Please*, Ann! And the radioactive dust, son?"

"An' if it was ra-di-o-ac-tive dust 'stead of atom bombs, my skin would come right off my body, an' my lungs would burn up inside me—please, poppa, I won't do it again!"

"And your eyes? What would happen to your eyes?"

A chubby brown fist dug into one of the eyes. "An' my eyes would fall out, an' my teeth would fall out, and I'd feel such terrible terrible pain—"

"All over and inside you. That's what would happen if you got to the cellar too late when the alarm went off, if you got locked out. At the end of three minutes, we pull the levers, and no matter who's outside—*no matter who*—all four corner doors swing shut and the cellar will be sealed. You understand that, Saul?"

The two Dawkins children were listening with white faces and dry lips. Their parents had brought them from the city and begged Elliot Plunkett as he remembered old friends to give their children the same protection as his. Well, they were getting it. This was the way to get it.

"Yes, I understand it, poppa. I won't ever do it again. Never again."

"I hope you won't. Now start for the barn, Saul. Go ahead." Plunkett slid his heavy leather belt from its loops.

"Elliot! Don't you think he understands the horrible thing? A beating won't make it any clearer."

He paused behind the weeping boy trudging to the barn. "It won't make it any clearer, but it will teach him the lesson another way. All seven of us are going to be in that cellar three minutes after the alarm, if I have to wear this strap clear down to the buckle!"

When Plunkett later clumped into the kitchen with his heavy farm boots, he stopped and sighed.

Ann was feeding Dinah. With her eyes on the baby, she asked, "No supper for him, Elliot?"

"No supper." He sighed again. "It does take it out of a man."

"Especially you. Not many men would become a farmer at thirty-five. Not many men would sink every last penny into an underground fort and powerhouse, just for insurance. But you're right."

"I only wish," he said restlessly, "that I could work out some way of getting Nancy's heifer into the cellar. And if eggs stay high one more month I can build the tunnel to the generator. Then, there's the well. Only one well, even if it's enclosed—"

"And when we came out here seven years ago—" She rose to him at last and rubbed her lips gently against his thick blue shirt. "We only had a piece of ground. Now, we have three chicken houses, a thousand broilers, and I can't keep track of how many layers and breeders."

She stopped as his body tightened and he gripped her shoulders.

"Ann, *Ann!* If you think like that, you'll act like that! How can I expect the children to—Ann, what we have—all we have—is a five room cellar, concrete-lined, which we can seal in a few seconds, an enclosed well from a fairly deep underground stream, a windmill generator for power and a sunken oil-burner-driven generator for emergencies. We have supplies to carry us through, Geiger counters to detect radiation and lead-lined suits to move about in—afterwards. I've told you again and again that these things are our lifeboat, and the farm is just a sinking ship.

"Of course, darling." Plunkett's teeth ground together, then parted helplessly as his wife went back to feeding Dinah, the baby.

"You're perfectly right. Swallow, now, Dinah. Why, that last bulletin from the Survivors Club would make *anybody* think."

He had been quoting from the October *Survivor* and Ann had recognized it. Well? At least they were *doing* something—seeking out nooks and feverishly building crannies—pooling their various ingenuities in an attempt to haul themselves and their families through the military years of the Atomic Age.

The familiar green cover of the mimeographed magazine was very noticeable on the kitchen table. He flipped the sheets to the thumb-smudged article on page five and shook his head.

"Imagine!" he said loudly. "The poor fools agreeing with the government again on the safety factor. Six minutes! How can they—an organization like the Survivors Club making that their official opinion! Why freeze, *freeze* alone. . . ."

"They're ridiculous," Ann murmured, scraping the bottom of the bowl.

"All right, we have automatic detectors. But human beings still have to look at the radar scope, or we'd be diving underground every time there's a meteor shower."

He strode along a huge table, beating a fist rhythmically into one hand. "They won't be so sure, at first. Who wants to risk his rank by giving the nationwide signal that makes everyone in the country pull ground over his head, that makes our own projectile sites set to buzz? Finally, they are certain: they freeze for a moment. Meanwhile, the rockets are zooming down—how fast, we don't know. The men unfreeze, they trip each other up, they tangle frantically. *Then* they press the button, *then* the nationwide signal starts our radio alarms."

Plunkett turned to his wife, spread earnest, quivering arms. "And then, Ann, *we* freeze when we hear it! At last, we start for the cellar. Who knows, who can dare to say, how much has been cut off the margin of safety by that time? No, if they claim that six minutes is the safety factor, we'll give half of it to the alarm system. Three minutes for us."

"One more spoonful," Ann urged Dinah. "Just one more. *Down* it goes!"

Josephine Dawkins and Herbie were cleaning the feed trolley in the shed at the near end of the chicken house.

"All done, pop," the boy grinned at his father. "And the eggs taken care of. When does Mr. Whiting pick 'em up?"

"Nine o'clock. Did you finish feeding the hens in the last house?"

"I said all done, didn't I?" Herbie asked with adolescent impatience. "When I say a thing, I mean it."

"Good. You kids better get at your books. Hey, stop that! Education will be very important, afterwards. You never know what will be useful. And maybe only your mother and I to teach you."

"Gee," Herbie nodded at Josephine. "Think of that."

She pulled at her jumper where it was very tight over newly swelling breasts and patted her blonde braided hair. "What about *my* mother and father, Mr. Plunkett? Won't they be—be—"

"Naw!" Herbie laughed the loud, country laugh he'd been practicing lately. "They're dead-enders. They won't pull through. They live in the City, don't they? They'll just be some—"

"Herbie!"

"—some foam on a mushroom-shaped cloud," he finished, utterly entranced by the image. "Gosh, I'm sorry," he said, as he looked from his angry father to the quivering girl. He went on in a studiously reasonable voice. "But it's the truth, anyway. That's why they sent you and Lester here. I guess I'll marry you—afterwards. And you ought to get in the habit of calling *him* pop. Because that's the way it'll be."

Josephine squeezed her eyes shut, kicked the shed door open, and ran out. "I hate you, Herbie Plunkett," she wept. "You're a beast!"

Herbie grimaced at his father—*women, women, women!*—and ran after her. "Hey, Jo! Listen!"

The trouble was, Plunkett thought worriedly as he carried the emergency bulbs for the hydroponic garden into the cellar

Generation of Noah

—the trouble was that Herbie had learned through constant reiteration the one thing: survival came before all else, and amenities were merely amenities.

Strength and self-sufficiency—Plunkett had worked out the virtues his children needed years ago, sitting in air-conditioned offices and totting corporation balances with one eye always on the calendar.

"Still," Plunkett muttered, "still—Herbie shouldn't—" He shook his head.

He inspected the incubators near the long steaming tables of the hydroponic garden. A tray about ready to hatch. They'd have to start assembling eggs to replace it in the morning. He paused in the third room, filled a gap in the bookshelves.

"Hope Josephine steadies the boy in his schoolwork. If he fails that next exam, they'll make me send him to town regularly. Now *there's* an aspect of survival I can hit Herbie with."

He realized he'd been talking to himself, a habit he'd been combating futilely for more than a month. Stuffy talk, too. He was becoming like those people who left tracts on trolley cars.

"Have to start watching myself," he commented. "Dammit, again!"

The telephone clattered upstairs. He heard Ann walk across to it, that serene, unhurried walk all pregnant women seem to have.

"Elliot! Nat Medarie."

"Tell him I'm coming, Ann." He swung the vault-like door carefully shut behind him, looked at it for a moment, and started up the high stone steps.

"Hello, Nat. What's new?"

"Hi, Plunk. Just got a postcard from Fitzgerald. Remember him? The abandoned silver mine in Montana? Yeah. He says we've got to go on the basis that lithium and hydrogen bombs will be used."

Plunkett leaned against the wall with his elbow. He cradled the receiver on his right shoulder so he could light a cigarette. "Fitzgerald can be wrong sometimes."

"Uhm. I don't know. But you know what a lithium bomb means, don't you?"

"It means," Plunkett said, staring through the wall of the house and into a boiling Earth, "that a chain reaction may be set off in the atmosphere if enough of them are used. Maybe if only one—"

"Oh, can it," Medarie interrupted. "That gets us nowhere. That way nobody gets through, and we might as well start shuttling from church to bar-room like my brother-in-law in Chicago is doing right now. Fred, I used to say to him— No, listen Plunk: it means I was right. You didn't dig deep enough."

"*Deep* enough! I'm as far down as I want to go. If I don't have enough layers of lead and concrete to shield me—well, if they can crack *my* shell, then you won't be able to walk on the surface before you die of thirst, Nat. No—I sunk my dough in power supply. Once that fails, you'll find yourself putting the used air back into your empty oxygen tanks by hand!"

The other man chuckled. "All right. I *hope* I see you around."

"And I hope *I* see . . ." Plunkett twisted around to face the front window as an old station wagon bumped over the ruts in his driveway. "Say, Nat, what do you know? Charlie Whiting just drove up. Isn't this Sunday?"

"Yeah. He hit my place early, too. Some sort of political meeting in town and he wants to make it. It's not enough that the striped-pants brigade are practically glaring into each other's eyebrows this time. A couple of local philosophers are impatient with the slow pace at which their extinction is approaching, and they're getting to see if they can't hurry it up some."

"Don't be bitter," Plunkett smiled.

"Here's praying at you. Regards to Ann, Plunk."

Plunkett cradled the receiver and ambled downstairs. Outside, he watched Charlie Whiting pull the door of the station wagon open on its one desperate hinge.

"Eggs stowed, Mr. Plunkett," Charlie said. "Receipt signed. Here. You'll get a check Wednesday."

"Thanks, Charlie. Hey, you kids get back to your books. Go on, Herbie. You're having an English quiz tonight. Eggs still going up, Charlie?"

"Up she goes." The old man slid onto the crackled leather seat and pulled the door shut deftly. He bent his arm on the open window. "Heh. And every time she does I make a little more off you survivor fellas who are too scairt to carry 'em into town yourself."

"Well, you're entitled to it," Plunkett said, uncomfortably. "What about this meeting in town?"

"Bunch of folks goin' to discuss the conference. I say we pull out. I say we walk right out of the dern thing. This country never won a conference yet. A million conferences the last few years and everyone knows what's gonna happen sooner or later. Heh. They're just wastin' time. Hit 'em first, I say."

"Maybe we will. Maybe *they* will. Or—maybe, Charlie—a couple of different nations will get what looks like a good idea at the same time."

Charlie Whiting shoved his foot down and ground the starter. "You don't make sense. If we hit 'em first how can they do the same to us? Hit 'em first—hard enough—and they'll never recover in time to hit us back. That's what *I* say. But you survivor fellas—" He shook his white head angrily as the car shot away.

"Hey!" he yelled, turning into the road. "Hey, look!"

Plunkett looked over his shoulder. Charlie Whiting was gesturing at him with his left hand, the forefinger pointing out and the thumb up straight.

"Look, Mr. Plunkett," the old man called. "Boom! Boom! Boom!" He cackled hysterically and writhed over the steering wheel.

Rusty scuttled around the side of the house, and after him, yipping frantically in ancient canine tradition.

Plunkett watched the receding car until it swept around the curve two miles away. He stared at the small dog returning proudly.

Poor Whiting. Poor everybody, for that matter, who had a normal distrust of crackpots.

How could you permit a greedy old codger like Whiting to buy your produce, just so you and your family wouldn't have to risk trips into town?

Well, it was a matter of having decided years ago that the world was too full of people who were convinced that they were faster on the draw than anyone else—and the other fellow was bluffing anyway. People who believed that two small boys could pile up snowballs across the street from each other and go home without having used them, people who discussed the merits of concrete fences as opposed to wire guard-rails while their automobiles skidded over the cliff. People who were righteous. People who were apathetic.

It was the last group, Plunkett remembered, who had made him stop buttonholing his fellows, at last. You got tired of standing around in a hair shirt and pointing ominously at the heavens. You got to the point where you wished the human race well, but you wanted to pull you and yours out of the way of its tantrums. Survival for the individual and his family, you thought—

Clang-ng-ng-ng-ng!

Plunkett pressed the stud on his stopwatch. Funny. There was no practice alarm schedule for today. All the kids were out of the house, except Saul—and he wouldn't dare to leave his room, let alone tamper with the alarm. Unless, perhaps, Ann—

He walked inside the kitchen. Ann was running toward the door, carrying Dinah. Her face was oddly unfamiliar. "Saulie!" she screamed. "Saulie! Hurry *up,* Saulie!"

"I'm coming, momma," the boy yelled as he clattered down the stairs. "I'm coming as fast as I can! I'll make it!"

Plunkett understood. He put a heavy hand on the wall, under the dinner-plate clock.

He watched his wife struggle down the steps into the cellar. Saul ran past him and out of the door, arms flailing. "I'll make it, poppa! I'll make it!"

Plunkett felt his stomach move. He swallowed with great care. "Don't hurry, son," he whispered. "It's only judgment day."

He straightened out and looked at his watch, noticing that his hand on the wall had left its moist outline behind. One minute, twelve seconds. Not bad. Not bad at all. He'd figured on three.

Clang-ng-ng-ng-ng!

He started to shake himself and began a shudder that he couldn't control. What was the matter? He knew what he had to do. He had to unpack the portable lathe that was still in the barn.

"Elliot!" his wife called.

He found himself sliding down the steps on feet that somehow wouldn't lift when he wanted them to. He stumbled through the open cellar door. Frightened faces dotted the room in an unrecognizable jumble.

"We all here?" he croaked.

"All here, poppa," Saul said from his position near the aeration machinery. "Lester and Herbie are in the far room, by the other switch. Why is Josephine crying? Lester isn't crying. I'm not crying, either."

Plunkett nodded vaguely at the slim, sobbing girl and put his hand on the lever protruding from the concrete wall. He glanced at his watch again. Two minutes, ten seconds. Not bad.

"Mr. Plunkett!" Lester Dawkins sped in from the corridor. "Mr. Plunkett! Herbie ran out of the other door to get Rusty. I told him—"

Two minutes, twenty seconds, Plunkett realized as he leaped to the top of the steps. Herbie was running across the vegetable garden, snapping his fingers behind him to lure Rusty on. When he saw his father, his mouth stiffened with shock. He broke stride for a moment, and the dog charged joyously between his legs. Herbie fell.

Plunkett stepped forward. *Two minutes, forty seconds.* Herbie jerked himself to his feet, put his head down—and ran.

Was that dim thump a distant explosion? There—another one! Like a giant belching. Who had started it? And did it matter—now?

Three minutes. Rusty scampered down the cellar steps, his head back, his tail flickering from side to side. Herbie panted up. Plunkett grabbed him by the collar and jumped.

And as he jumped he saw—far to the south—the umbrellas opening their agony upon the land. Rows upon swirling rows of them. . . .

He tossed the boy ahead when he landed. *Three minutes, five seconds.* He threw the switch, and, without waiting for the door to close and seal, darted into the corridor. That took care of two doors; the other switch controlled the remaining entrances. He reached it. He pulled it. He looked at his watch. *Three minutes, twenty seconds.* "The bombs," blubbered Josephine. "The bombs!"

Ann was scrabbling Herbie to her in the main room, feeling his arms, caressing his hair, pulling him in for a wild hug and crying out yet again. "Herbie! Herbie! Herbie!"

"I know you're gonna lick me, pop. I—I just want you to know that I think you ought to."

"I'm not going to lick you, son."

"You're not? But gee, I deserve a licking. I deserve the worst—"

"You may," Plunkett said, gasping at the wall of clicking geigers. *"You may deserve a beating,"* he yelled, so loudly that they all whirled to face him, "but I won't punish you, not only for now, but forever! And as I with you," he screamed, "so you with yours! Understand?"

"Yes," they replied in a weeping, ragged chorus. "We understand!"

"Swear! Swear that you and your children and your children's children will never punish another human being—*no matter what the provocation.*"

"We swear!" they bawled at him. "We swear!"

Then they all sat down.

To wait.

THE RATS

By Arthur Porges

He cuddled the stock against his shoulder, lined up the ivory bead, and squeezed the trigger. He heard the smack of the hollow-point against wood, and swore, his imprecations echoing hollowly down the dark, empty streets.

Jeffrey Clark expected no reply to his oaths, and got none. The silent village had been evacuated months before because of dangerous radio-activity from the adjoining proving ground for atomic weapons, now also abandoned.

Clark was a physicist, and understood perfectly that the government could not take chances. He knew that present radiation was quite harmless a short distance from the firing range, and there were excellent reasons for remaining here after the jerry-built settlement was evacuated.

In this region, wasteland to begin with, and now forbidden by law, a man would be safe. What enemy, he reasoned, cared to waste a gram of fissionable material on such a locality? Further, when the bombs fell, an eventuality he believed imminent, there would be no panicky mobs to pillage his supplies, menace his life blindly, and, in short, ruin his slender chance for survival.

There was a large store of food in his house, carefully built up during the three-year period when he worked on the proving ground. A small spring provided the only dependable supply of water within hundreds of square miles of desert; the government had left behind dozens of large drums of gasoline, as well as tons of miscellaneous equipment; and Clark was tough enough psychologically to make a good fight of it alone.

The only annoyance—at present, he rated it no higher—was the rats. With the abandonment of the village, they had found themselves short of food. Unable to follow the inhabitants across a pitiless desert, they were in a hopeless predicament. How they had arrived in the first place was a minor

---
Copyright, 1950, by Lock Publishing Corp., and originally published in *Man's World*, February, 1951.

mystery to the physicist, but he surmised that a few pairs had been hidden in the huge shipments of material; certainly the once-numerous mice, now almost exterminated by their large cousins, had come that way.

In any case, Clark was more interested in their future than their past, for he was finding it difficult to protect his possessions, especially the priceless food, against their inroads.

True, he had the .22 rifle and a large quantity of high-power shells, but the rats were no longer easy targets. Strange as it seemed to him at first, he was convinced they had learned to duck at the flash, like veteran infantrymen. He often bemoaned his blindness in failing to provide rattraps, but it was too late, now. Any contact with the outer world was definitely taboo. He had no wish either to share his retreat or to be conscripted for the Armageddon which lay a few calendar leaves ahead.

Blowing into the chamber of his gun, Clark returned moodily to the house. Something had to be done. With all his skill, he hadn't shot a rat in days, and the big albino that had just escaped had certainly been a perfect target. And in spite of the food scarcity, they still swarmed in great numbers throughout the town. Either they fed on each other, or else they had learned to catch the ubiquitous lizards that crouched with gently-throbbing throats on every sunny surface.

"The question is," he muttered, filling a pipe, "do I concentrate on purely defensive measures, like rat-proofing this house, or take the offensive?" He had repeatedly plugged ratholes with a mixture of cement and powdered glass, a mortar no rat cared to gnaw for long, but the enemy merely made new passages in the wooden dwelling.

Then, too, a number of minor incidents tended to make Clark uneasy. There was the hole, for example, which he had blocked with a sheet of tin. To his astonishment the rats had managed to strip away the metal, not by haphazard attacks, but through working directly on the tack-heads. Clark was no biologist, but he felt sure such intelligence was uncommon. For a moment he thought, idiotically enough, of a typical professional article: "Observations on an Unusually Adaptable Colony of—" what was the scientific name of the rat, again? *"Mus"* something. As if that mattered, with a world waiting for the assassin's blade. Anyhow, the explanation surely lay in the critical plight of the cunning rodents.

As he sat in the clear white glow of the gasoline lamp, puffing thoughtfully on his briar, a memory of childhood came to him, and he sat bolt upright.

"By George!" he exclaimed. "I should have thought of that

sooner. Gramps used to get hundreds that way, out on the farm."

Filled with enthusiasm, he decided to begin at once, although it was late. Living alone, he cared little about clock time, preferring a more flexible, wholly subjective measure.

A brief search located a large, empty barrel, which he sunk in the ground for about two-thirds its depth. This he filled half full of water from the nearby spring. Then, with great care, he adjusted a long plank so that it led rampwise from the ground, over the barrel's rim, to a point directly above the water. A few well-placed nails kept the board from moving laterally, while permitting free motion vertically, see-saw fashion. By repeated trials, he arranged the plank so that even a small weight in addition to the bait would destroy the delicate balance, sharply dipping the upper end.

After some moments of self-debate, in which he tried to brighten a dim memory of childhood, he placed several rocks in the water so as to form a tiny island.

Then, with a grunt of satisfaction, he fastened some scraps of food to the high end, and, well-pleased, returned to his house.

The moon was shining with that metallic brightness so typical of the clear desert air, and in a highly anticipatory mood Clark seated himself by the window with a 7 x 50 nightglass in his hand. He had not long to wait. Almost immediately he could see through the powerful lenses a group of lithe, furtive forms converging on the barrel with its promise of food. A leading rat, after hesitating briefly on the lower edge of the ramp, crept cautiously towards the top. It had just reached the bait, and was about to attack it with savage hunger, when the balance shifted; the plank dipped in one swift motion, and with a despairing squeak the rodent was plunged into chilly water.

For several moments as it swam about, clawing vainly at the smooth sides and squealing its indignation, the other rats vanished, but when the victim scrambled aboard the rock island and continued to shriek for help, they quickly reassembled, drawn by irresistible curiosity.

To their surprise, the mysterious pathway had returned to its original position, lying invitingly before them. Their natural desire for food was supplemented now by a burning urge to know what was happening to their fellow, still keening loudly, but invisible; and before long a second rodent attempted the incline.

Clark roared with laughter as the board, working with the

The Rats

simple efficiency of perfect design, dropped a second rat into cold water.

Both rats were squealing now, long reedy cries of fear and rage. With diabolical intent, Clark had made the island large enough for only one rat, and a grim battle for possession began.

Excited by the cries, and unable to see what was happening, the free rats returned in hordes, and utterly reckless in their madness, dashed up the treacherous ramp. Only a few held back, among them the large albino, and before dawn Clark's barrel-trap had swallowed fifteen rats, a record it maintained throughout the week.

It was on the tenth day that the situation changed.

Watching through his binoculars, Clark saw a rat hesitate on the lower edge, as usual. Another, close behind, impatiently shouldered by, quickly reaching the top, with its odorous bait. As the first rat still paused irresolutely below, the more daring one actually reached the food, tearing at it ravenously. This sight proved too much for the timid one, and it jealously rushed to join in the feast. With a double weight towards the top, the plank immediately hurled both animals to a watery death. Clark laughed at this byplay until his sides ached. The rats were so human in their reactions. Or should that be put in reverse, he wondered?

But ten minutes later something happened that wiped the smile from his lips. This time the albino took a hand, remaining calmly on the lower edge while a companion raced up the incline. At the top, the rat tore loose a large fragment of rancid bacon, and beat a nervous retreat. Clark could have sworn the animal looked positively relieved on reaching the ground again.

"Well, I'm damned!" Clark muttered. "Was that intentional or—!"

A few more nights' watching answered that question, and the barrel claimed no more victims.

Although concerned by this setback, Clark was far from beaten. If traps—or at least this type—were futile, there still remained other methods. Poison, for example. An inventory of his supplies, however, proved discouraging. Beyond a small stock of medical drugs, there was not a grain of poison to be had. He made a few tentative trials with ground glass, but found, as a toxicologist had once insisted, that it was nearly harmless.

No, poison in the ordinary sense was out, but death by swallowing didn't necessarily mean chemicals or glass. Clark was thinking of a device often used by Eskimos against bears and foxes. It was simple and effective. You coiled a thin sliver of whalebone into a tight, small spiral, and froze it in a pellet of fat. When an animal swallowed such a lump, it soon thawed out; the deadly coil snapped open, and the sharp-pointed bone pierced the creature's vitals.

Of course, he wouldn't use whalebone, nor was freezing called for. Clark rummaged about in the miscellaneous supplies and found some stiff, springy wire. He cut it into three-inch lengths, which he wound, under heavy tension, to spirals no larger than beans. He made a quantity of such coils, all tied with thread. There wasn't much doubt, he decided, considering what rats ate, that the thread would quickly weaken in their digestive fluids. Then, bingo!

The results were heartening beyond his expectations. Concealed in pills of stale food, or small lumps of flour paste, the murderous spirals soon disposed of several dozen rats, and Clark began to hope that total extinction was possible.

On that score he was soon undeceived. These rats learned with amazing rapidity, and before long the pills mouldered away uneaten where he left them.

Meanwhile the creatures were bolder than ever. One night, after filling his dish with food, Clark stepped into the pantry for some salt. Almost immediately he heard a scuffling noise in the outer room, and feeling certain a rat was after his dinner, sprang out just in time to see a slinking white form slip oil-like under a heavy bookcase. It was the albino again, apparently a leader among the rodents. Clark angrily muscled the massive case aside, and sure enough, a freshly-gnawed passage gaped in the corner behind it. Swearing, he returned to the table, where as he ate, ideas for a new, intensive campaign were mentally marshalled and analyzed.

He was chewing a mouthful thoughtfully before swallowing it when his teeth grated on metal. He paled, fighting sudden almost overwhelming nausea. Then, very carefully, with fingers that shook, he removed from the back of his tongue a shattered pill of hard biscuit. Most of the shell had been stripped off by the action of teeth and moisture, leaving the terrible little spiral plainly visible.

Clark shuddered. But for the accident of teeth meeting metal, the small pellet might have been swallowed. There was no doctor for a hundred miles, and with three inches of sharp wire jammed into stomach or bowels—well, no rat could be more hopelessly doomed.

But that wasn't the point, now. How did the damned thing

get on his plate? He had been extremely careful not to leave the tricky pills about. A man living alone learns to take every precaution against accidents of all sorts. Then he remembered the white rat. But that was absurd. Surely it hadn't deliberately dropped the pellet into his food. Rats were adaptable, and these exceptionally so, but this sort of human reasoning was as far beyond them as building a railroad.

When he had recovered his composure, Clark inspected the remaining food minutely, but there were no more spirals. Nevertheless his appetite was gone, and leaving the table, he dropped into a chair, there to puff pensively on a pipe.

"If I only had a cat," he murmured, thinking longingly of the mighty, sandy Tom of his childhood. "Cap'n Kidd would make short work of this lousy vermin."

But it was useless to think of cats; action was called for, and quickly. What he needed, Clark felt, was a large, efficient trap that would shatter the whole rat colony at one blow. After that, cleaning up a few survivors might be possible before rodent fertility made good their losses.

There was a small, sturdy shed a hundred feet from the house, and Clark decided to use that. A careful inspection proved it to be eminently suitable, but just as a precaution, he reinforced it with boards here and there, stopped up a few rat holes, and placed tin sheathing at strategic points.

It was simple enough to build a heavy door that could be released from the house by a cord. He had it slide vertically in oiled grooves, dropping smoothly with great speed. Of course, it was a bit large, but that was no problem, and made inspection of the interior easy.

He wondered about a catch, but concluded no rat in the world could budge the weighty door once it fell. Still, these were remarkable animals, and he ought to play safe. After all, if this trap failed, there wasn't much more to try. An automatic lock was uncalled for, but there was no harm in having a pair of staples at the bottom, and a short stick to engage them. Not that the rats would have time to do much with all that dry wood piled about the shed ready for his match.

When everything was set, he placed a quantity of spoiled food in the shed, and returned to the house. It would take several days, he knew, before the harried rats would enter the suspicious structure freely, but their actual precautions were a revelation. Having succumbed in large numbers to the wire pills, the rats were unbelievably wary. From his window, Clark watched through binoculars, and for three days, as the animals came to the shed in dozens, he marvelled at their latest procedure.

Apparently a small group of the rodents were tasters, since before mass feeding began, they scouted the food piles, nibbling everywhere with excellent sampling technique. Only when these potential martyrs remained unharmed for a reasonable period of time, did the main body approach.

But tasters or not, they entered the shed, and by the fifth day in such hordes that Clark felt certain there were few holdouts.

During the late afternoon a week later, therefore, he made his final preparations, replenishing the food, adjusting the cord, and testing the sliding door. He was about to leave, well-satisfied, when sudden doubt assailed him. Had he overlooked something? Yes, by George. Suppose the cunning rats had outwitted him by digging a few secret bolt-holes recently. What a fool he'd be, if after all this trouble he fired the locked shed only to have rats pour out of a dozen new holes. True, there was tin along most of the lower wall, and the floor was concrete, but with these rats it was best to make sure.

Stooping, he re-entered the shed, and began a painstaking examination of each metal sheet. While he was fingering the nail-heads, he heard a shuffling noise outside, accompanied by loud squeaks. He smiled sourly. The victims were already gathering for their last feast. The sounds grew louder; they came from the roof, too. He decided to step out and check up. The cord passed through a pulley there, and some rat might jam it—he was in a mood to believe they might do so intentionally, even if that seemed fantastic.

He had taken only one step towards the door, however, when it fell with a crash. Clark stopped in his tracks, swearing angrily. How had that happened? The catch was smooth-working, but still needed a reasonably hard tug on its cord. A hint of panic touched him. Could the rats have done the trapping? No, that was insane. Yet, if they could hold him here for even an hour, with his food unguarded—listen, they were at the door now. Well, he was no damned rat. One yank at the oiled door, and he'd be free. He dug his nails into the rough wood and tugged. The door rose smoothly half an inch, then stopped dead. Perspiration burned his eyes. He exerted all his strength. No dice. It was jammed, all right. He put one eye to a crack, trying to locate the trouble, and saw the great albino just outside. Raging, he peered through several slits before understanding. The short, thick dowel-rod he'd brought to engage the staples was neatly in place. The rats had locked him in. They were all about, and surely there was obscene triumph in their scurryings and squealings.

It was obvious he had completely underestimated them. Yet they had a lot to learn, he thought grimly, regaining his poise. This shed couldn't hold a man very long. He pulled out a heavy pocketknife, hesitated, then returned it, and instead, drew the long-barreled .22 automatic from his belt. A few well placed shots would splinter the door enough to let him reach that dowel-rod. He peered out again to locate the best point, and in the growing gloom saw a bobbing light, then another, and a third. For a moment he thought wonderingly of human aid, but these lights were almost at ground level. Then his heart pounded, and he saw all too plainly. They were rats, each with a flaming stick in its jaws. There was only one explanation now, that was certain. The sticks had been lit at his own gas lamp burning at home, and the motive was horribly clear.

Cursing, half sobbing, he battered frantically at the thick wood. He fired until the gun was empty, but the light slugs only chewed up the door's surface, and in the ensuing silence he heard the crackling flames on three sides.

Abruptly he was calm, and the whole situation seemed humorously ironical as full comprehension came. There were not just highly adaptable rats. Everybody knew that radiation did strange things to living cells, and these creatures had been long exposed. No, they were no more rats than men were apes. These were intelligent, quick-learning mutants, and the huge albino was surely their leader.

Clark felt coolly in his pockets. Yes, a break at last. One bullet left. The heat was stifling; there wasn't much time. He raised the loaded gun to his temple, and above the roaring flames heard a detestable, reedy keening.

At that moment as he stood poised between life and death, there was a flash of light somewhere over the horizon, transient yet so intense the very walls of the shed seemed transparent. The ground quivered faintly, as if a premonitory shiver was running over the world, and far off rumblings sounded threateningly.

Penned up though he was, the physicist understood perfectly. Without being able to see it, he knew the inevitable mushroom was having its brief flowering, tall and sinister, yet a thing of urgent beauty to the dispassionate observer.

Clark sobbed dryly. The raw fibre of his brain was touched with acid. Twice his lips moved soundlessly, stickily, before he said softly, "That was It."

The heat was now utterly unbearable, and even the fate of a world was secondary. He raised his voice to a shout, addressing the squealing mutants outside.

"You out there!" he roared, cringing from the searching flames. "You win, damn you! You may be the only ones left this time next month. It's all yours now. And what the hell will *you* do with it?" Then he squeezed the trigger.

ARARAT

By Zenna Henderson

WE'VE had trouble with teachers in Cougar Canyon. It's just an Accommodation school anyway, isolated and so unhandy to anything. There's really nothing to hold a teacher. But the way The People bring forth their young, in quantities and with regularity, even our small Group can usually muster the nine necessary for the County School Superintendent to arrange for the schooling for the year.

Of course I'm past school age, Canyon school age, and have been for years, but if the tally came up one short in the Fall, I'd go back for a post-graduate course again. But now I'm working on a college level because Father finished me off for my high school diploma two summers ago. He's promised me that if I do well this year I'll get to go Outside next year and get my training and degree so I can be the teacher and we won't have to go Outside for one any more. Most of the kids would just as soon skip school as not, but the Old Ones don't hold with ignorance and the Old Ones have the last say around here.

Father is the head of the school board. That's how I get in on lots of school things the other kids don't. This summer when he wrote to the County Seat that we'd have more than our nine again this fall and would they find a teacher for us, he got back a letter saying they had exhausted their supply of teachers who hadn't heard of Cougar Canyon and we'd have to dig up our own teacher this year. That "dig up" sounded like a dirty crack to me since we have the graves of four past teachers in the far corner of our cemetery. They sent us such old teachers, the homeless, the tottering, who were trying to piece out the end of their lives with a year here and a year there in jobs no one else wanted because there's no adequate pension system in the state and most teachers seem to die in harness. And their oldness and their tottering were not suffi-

Copyright, 1952, by Fantasy House, Inc., and originally published in *The Magazine of Fantasy and Science Fiction*, October, 1952.

cient in the Canyon where there are apt to be shocks for Outsiders—unintentional as most of them are.

We haven't done so badly the last few years, though. The Old Ones say we're getting adjusted—though some of the non-conformists say that The Crossing thinned our blood. It might be either or both or the teachers are just getting tougher. The last two managed to last until just before the year ended. Father took them in as far as Kerry Canyon and ambulances took them on in. But they were all right after a while in the sanatorium and they're doing okay now. Before them, though, we usually had four teachers a year.

Anyway, Father wrote to a Teachers Agency on the coast and after several letters each way, he finally found a teacher.

He told us about it at the supper table.

"She's rather young," he said, reaching for a toothpick and tipping his chair back on its hind legs.

Mother gave Jethro another helping of pie and picked up her own fork again. "Youth is no crime," she said, "and it'll be a pleasant change for the children."

"Yes, though it seems a shame." Father prodded at a back tooth and Mother frowned at him. I wasn't sure if it was for picking his teeth or for what he said. I knew he meant it seemed a shame to get a place like Cougar Canyon so early in a career. It isn't that we're mean or cruel, you understand. It's only that they're Outsiders and we sometimes forget—especially the kids.

"She doesn't *have* to come," said Mother. "She could say no."

"Well, now—" Father tipped his chair forward. "Jethro, no more pie. You go on out and help 'Kiah bring in the wood. Karen, you and Lizbeth get started on the dishes. Hop to it, kids."

And we hopped, too. Kids do to fathers in the Canyon, though I understand they don't always Outside. It annoyed me because I knew Father wanted us out of the way so he could talk adult talk to Mother, so I told Lizbeth I'd clear the table and then worked as slowly as I could, and as quietly, listening hard.

"She couldn't get any other job," said Father. "The agency told me they had placed her twice in the last two years and she didn't finish the year either place."

"Well," said Mother, pinching in her mouth and frowning. "If she's that bad, why on earth did you hire her for the Canyon?"

"We have a choice?" laughed Father. Then he sobered. "No,

it wasn't for incompetency. She was a good teacher. The way she tells it, they just fired her out of a clear sky. She asked for recommendations and one place wrote, 'Miss Carmody is a very competent teacher but we dare not recommend her for a teaching position.' "

" 'Dare not'?" asked Mother.

" 'Dare not,' " said Father. "The Agency assured me that they had investigated thoroughly and couldn't find any valid reasons for the dismissals, but she can't seem to find another job anywhere on the coast. She wrote me that she wanted to try another state."

"Do you suppose she's disfigured or deformed?" suggested Mother.

"Not from the neck up!" laughed Father. He took an envelope from his pocket. "Here's her application picture."

By this time I'd got the table cleared and I leaned over Father's shoulder.

"Gee!" I said. Father looked back at me, raising one eyebrow. I knew then that he had known all along that I was listening.

I flushed but stood my ground, knowing I was being granted admission to adult affairs, if only by the back door.

The girl in the picture was lovely. She couldn't have been many years older than I and she was twice as pretty. She had short dark hair curled all over her head and apparently that poreless creamy skin that seems to have an inner light of itself. She had a tentative look about her as though her dark eyebrows were horizontal question marks. There was a droop to the corners of her mouth—not much, just enough to make you wonder why . . . and want to comfort her.

"She'll stir the Canyon for sure," said Father.

"I don't know," Mother frowned thoughtfully. "What will the Old Ones to say to a marriageable Outsider in the Canyon?"

"Adonday Veeah!" muttered Father. "That never occurred to me. None of our other teachers were ever of an age to worry about."

"What *would* happen?" I asked. "I mean if one of The Group married an Outsider?"

"Impossible," said Father, so like the Old Ones that I could see why his name was approved in Meeting last Spring.

"Why, there's even our Jemmy," worried Mother. "Already he's saying he'll have to start trying to find another Group. None of the girls here please him. Supposing this Outsider— how old is she?"

Father unfolded the application. "Twenty-three," he said, "Just three years out of college."

"Jemmy's twenty-four," said Mother, pinching her mouth together. "Father, I'm afraid you'll have to cancel the contract. If anything happened— Well, you waited over-long to become an Old One to my way of thinking and it'd be a shame to have something go wrong your first year."

"I can't cancel the contract. She's on her way here. School starts next Monday." Father ruffled his hair forward as he does when he's disturbed. "We're probably making a something of a nothing," he said hopefully.

"Well, I only hope we don't have any trouble with this Outsider."

"Or she with us," grinned Father. "Where are my cigarettes?"

"On the book case," said Mother, getting up and folding the table cloth together to hold the crumbs.

Father snapped his fingers and the cigarettes drifted in from the front room.

Mother went on out to the kitchen. The table cloth shook itself over the waste basket and then followed her.

Father drove to Kerry Canyon Sunday night to pick up our new teacher. She was supposed to have arrived Saturday afternoon, but she didn't make bus connections at the County Seat. The road ends at Kerry Canyon. I mean for Outsiders. There's not much of the look of a well-traveled road very far out our way from Kerry Canyon, which is just as well. Tourists leave us alone. Of course *we* don't have much trouble getting our cars to and fro but that's why everything dead-ends at Kerry Canyon and we have to do all our own fetching and carrying —I mean the road being in the condition it is.

All the kids at our house wanted to stay up to see the new teacher, so Mother let them, but by 7:30 the youngest ones began to drop off and by 9 there was only Jethro and 'Kiah, Lizbeth and Jemmy and me. Father should have been home long before and Mother was restless and uneasy. I knew if he didn't arrive soon, she would head for her room and the cedar box under the bed. But at 9:15 we heard the car coughing and sneezing up the draw. Mother's wide relieved smile was reflected on all our faces.

"Of course!" she cried. "I forgot. He has an Outsider in the car. He had to use the *road* and it's terrible across Jackass Flat."

I felt Miss Carmody before she came in the door. I was tingling all over from anticipation already, but all at once I felt her, so plainly that I knew with a feeling of fear and pride that I was of my Grandmother, that soon I would be bearing the burden and blessing of her Gift: the Gift that develops into free access to any mind—one of The People or Outsider—willing or not. And besides the access, the ability to council and help, to straighten tangled minds and snarled emotions.

And then Miss Carmody stood in the doorway, blinking a little against the light, muffled to the chin against the brisk fall air. A bright scarf hid her hair but her skin *was* that luminous matt-cream it had looked. She was smiling a little, but scared, too. I shut my eyes and . . . I went in—just like that. It was the first time I had ever sorted anybody. She was all fluttery with tiredness and strangeness and there was a question deep inside her that had the wornness of repetition, but I couldn't catch what it was. And under the uncertainty there was a sweetness and dearness and such a bewildered sorrow that I felt my eyes dampen. Then I looked at her again (sorting takes such a little time) as Father introduced her. I heard a gasp beside me and suddenly I went into Jemmy's mind with a stunning rush.

Jemmy and I have been close all our lives and we don't always need words to talk with one another, but this was the first time I had ever gone in like this and I knew he didn't know what had happened. I felt embarrassed and ashamed to know his emotion so starkly. I closed him out as quickly as possible, but not before I knew that now Jemmy would never hunt for another Group; Old Ones or no Old Ones, he had found his love.

All this took less time than it takes to say "How do you do?" and shake hands. Mother descended with cries and drew Miss Carmody and Father out to the kitchen for coffee and Jemmy swatted Jethro and made him carry the luggage instead of snapping it to Miss Carmody's room. After all, we didn't want to lose our teacher before she even saw the school house.

I waited until everyone was bedded down. Miss Carmody in her cold, cold bed, the rest of us of course with our sheets set for warmth—how I pity Outsiders! Then I went to Mother.

She met me in the dark hall and we clung together as she comforted me.

"Oh Mother," I whispered. "I sorted Miss Carmody tonight. I'm afraid."

Mother held me tight again. "I wondered," she said. "It's a great responsibility. You have to be so wise and clear-thinking. Your Grandmother carried the Gift with graciousness and honor. You are of her. You can do it."

"But Mother! To be an Old One!"

Mother laughed. "You have years of training ahead of you before you'll be an Old One. Councilor to the soul is a weighty job."

"Do I have to tell?" I pleaded. "I don't want anyone to know yet. I don't want to be set apart."

"I'll tell the Oldest," she said, "No one else need know." She hugged me again and I went back, comforted, to bed.

I lay in the darkness and let my mind clear, not even knowing how I knew how to. Like the gentle reachings of quiet fingers I felt the family about me. I felt warm and comfortable as though I were cupped in the hollow palm of a loving hand. Some day I would belong to the Group as I now belonged to the family. Belong to others? With an odd feeling of panic, I shut the family out. I wanted to be alone—to belong just to me and no one else. I didn't *want* the Gift.

I slept after a while.

Miss Carmody left for the school house an hour before we did. She wanted to get things started a little before school time, her late arrival making it kind of rough on her. 'Kiah, Jethro, Lizbeth and I walked down the lane to the Armisters' to pick up their three kids. The sky was so blue you could taste it, a winey, fallish taste of harvest fields and falling leaves. We were all feeling full of bubbly enthusiasm for the beginning of school. We were light-hearted and light-footed, too, as we kicked along through the cottonwood leaves paving the lane with gold. In fact Jethro felt too light-footed and the third time I hauled him down and made him walk on the ground, I cuffed him good. He was still sniffling when we got to Armisters'.

"She's pretty!" called Lizbeth before the kids got out to the gate, all agog and eager for news of the new teacher.

"She's young," added 'Kiah, elbowing himself ahead of Lizbeth.

"She's littler'n me," sniffled Jethro and we all laughed because he's five-six already even if he isn't twelve yet.

Debra and Rachel Armister linked arms with Lizbeth and scuffled down the lane, heads together, absorbing the details of teacher's hair, dress, nail polish, luggage and night clothes, though goodness knows how Lizbeth knew anything about that.

Jethro and 'Kiah annexed Jeddy and they climbed up on the rail fence that parallels the lane and walked the top rail. Jethro took a tentative step or two above the rail, caught my eye and stepped back in a hurry. He knows as well as any child in the

Canyon that a kid his age has no business lifting along a public road.

We detoured at the Mesa Road to pick up the Kroginold boys. More than once Father has sighed over the Kroginolds.

You see, when The Crossing was made, The People got separated in that last wild moment when air was screaming past and the heat was building up so alarmingly. The members of our Group left their ship just seconds before it crashed so devastatingly into the box canyon behind Old Baldy and literally splashed and drove itself into the canyon walls, starting a fire that stripped the hills bare for miles. After The People gathered themselves together from the Life Slips and founded Cougar Canyon, they found that the alloy the ship was made of was a metal much wanted here. Our Group has lived on mining the box canyon ever since, though there's something complicated about marketing the stuff. It has to be shipped out of the country and shipped in again because everyone knows that it doesn't occur in this region.

Anyway, our Group at Cougar Canyon is probably the largest of The People, but we are reasonably sure that at least one Group and maybe two survived along with us. Grandmother in her time sensed two Groups but could never locate them exactly and, since our object is to go unnoticed in this new life, no real effort has ever been made to find them. Father can remember just a little of The Crossing, but some of the Old Ones are blind and crippled from the heat and the terrible effort they put forth to save the others from burning up like falling stars.

But getting back, Father often said that of all The People who could have made up our Group, we had to get the Kroginolds. They're rebels and were even before The Crossing. It's their kids that have been so rough on our teachers. The rest of us usually behave fairly decently and remember that we have to be careful around Outsiders.

Derek and Jake Kroginold were wrestling in a pile of leaves by the front gate when we got there. They didn't even hear us coming, so I leaned over and whacked the nearest rear-end and they turned in a flurry of leaves and grinned up at me for all the world like pictures of Pan in the mythology book at home.

"What kinda old bat we got this time?" asked Derek as he scrabbled in the leaves for his lunch box.

"She's not an old bat," I retorted, madder than need be because Derek annoys me so. "She's young and beautiful."

"Yeah, I'll bet!" Jake emptied the leaves from his cap onto the trio of squealing girls.

"She is so!" retorted 'Kiah. "The nicest teacher we ever had."

"She won't teach me nothing!" yelled Derek, lifting to the top of the cottonwood tree at the turn-off.

"Well, if she won't, I will," I muttered and, reaching for a handful of sun, I platted the twishers so quickly that Derek fell like a rock. He yelled like a catamount, thinking he'd get killed for sure, but I stopped him about a foot from the ground and then let go. Well, the stopping and the thump to the ground pretty well jarred the wind out of him, but he yelled:

"I'll tell the Old Ones! You ain't supposed to platt twishers—!"

"Tell the Old Ones," I snapped, kicking on down the leafy road. "I'll be there and tell them why. And then, old smarty pants, what will be your excuse for lifting?"

And then I was ashamed. I was showing off as bad as a Kroginold—but they make me so mad!

Our last stop before school was at the Clarinades'. My heart always squeezed when I thought of the Clarinade twins. They just started school this year—two years behind the average Canyon kid. Mrs. Kroginold used to say that the two of them, Susie and Jerry, divided one brain between them before they were born. That's unkind and untrue—thoroughly a Kroginold remark—but it is true that by Canyon standards the twins were retarded. They lacked so many of the attributes of The People. Father said it might be a delayed effect of The Crossing that they would grow out of, or it might be advance notice of what our children will be like here—what is ahead for The People. It makes me shiver, wondering.

Susie and Jerry were waiting, clinging to one another's hand as they always were. They were shy and withdrawn, but both were radiant because of starting school. Jerry, who did almost all the talking for the two of them, answered our greetings with a shy "Hello."

Then Susie surprised us all by exclaiming, "We're going to school!"

"Isn't it wonderful?" I replied, gathering her cold little hand into mine. "And you're going to have the prettiest teacher we ever had."

But Susie had retired into blushing confusion and didn't say another word all the way to school.

I was worried about Jake and Derek. They were walking apart from us, whispering, looking over at us and laughing.

They were cooking up some kind of mischief for Miss Carmody. And more than anything I wanted her to stay. I found right then that there *would* be years ahead of me before I became an Old One. I tried to go in to Derek and Jake to find out what was cooking, but try as I might I couldn't get past the sibilance of their snickers and the hard, flat brightness of their eyes.

We were turning off the road into the school yard when Jemmy, who should have been up at the mine long since, suddenly stepped out of the bushes in front of us, his hands behind him. He glared at Jake and Derek and then at the rest of the children.

"You kids mind your manners when you get to school," he snapped, scowling. "And you Kroginolds—just try anything funny and I'll lift you to Old Baldy and platt the twishers on you. This is one teacher we're going to keep."

Susie and Jerry clung together in speechless terror. The Kroginolds turned red and pushed out belligerent jaws. The rest of us just stared at a Jemmy who never raised his voice and never pushed his weight around.

"I mean it, Jake and Derek. You try getting out of line and the Old Ones will find a few answers they've been looking for —especially about the belfry in Kerry Canyon."

The Kroginolds exchanged looks of dismay and the girls sucked in breaths of astonishment. One of the most rigorously enforced rules of The Group concerns showing off outside the community. If Derek and Jake *had* been involved in ringing that bell all night last Fourth of July . . . well!

"Now you kids, scoot!" Jemmy jerked his head toward the schoolhouse and the terrified twins scudded down the leaf-strewn path like a pair of bright leaves themselves, followed by the rest of the children with the Kroginolds looking sullenly back over their shoulders and muttering.

Jemmy ducked his head and scowled. "It's time they got civilized anyway. There no sense to our losing teachers all the time."

"No," I said noncommittally.

"There's no point in scaring her to death," Jemmy was intent on the leaves he was kicking with one foot.

"No," I agreed, suppressing my smile.

Then Jemmy smiled ruefully in amusement at himself. "I should waste words with you," he said. "Here." He took his hands from behind him and thrust a bouquet of burning bright autumn leaves into my arms. "They're from you to her," he said. "Something pretty for the first day."

"Oh, Jemmy!" I cried through the scarlet and crimson and gold. "They're beautiful. You've been up on Baldy this morning."

"That's right," he said. "But she won't know where they came from." And he was gone.

I hurried to catch up with the children before they got to the door. Suddenly overcome with shyness, they were milling around the porch steps, each trying to hide behind the others.

"Oh, for goodness' sakes!" I whispered to our kids. "You ate breakfast with her this morning. She won't bite. Go on in."

But I found myself shouldered to the front and leading the subdued group into the school room. While I was giving the bouquet of leaves to Miss Carmody, the others with the ease of established habit slid into their usual seats, leaving only the twins, stricken and white, standing alone.

Miss Carmody, dropping the leaves on her desk, knelt quickly beside them, pried a hand of each gently free from their frenzy clutching and held them in hers.

"I'm so glad you came to school," she said in her warm, rich voice. "I need a first grade to make the school work out right and I have a seat that must have been built on purpose for twins."

And she led them over to the side of the room, close enough to the old pot-bellied stove for Outside comfort later and near enough to the window to see out. There, in dusted glory, stood one of the old double desks that The Group must have inherited from some ghost town out in the hills. There were two wooden boxes for footstools for small dangling feet and, spouting like a flame from the old ink well hole, a spray of vivid red leaves—matchmates to those Jemmy had given me.

The twins slid into the desk, never loosing hands, and stared up at Miss Carmody, wide-eyed. She smiled back at them and, leaning forward, poked her finger tip into the deep dimple in each round chin.

"Buried smiless," she said, and the two scared faces lighted up briefly with wavery smiles. Then Miss Carmody turned to the rest of us.

I never did hear her introductory words. I was too busy mulling over the spray of leaves, and how she came to know the identical routine, words and all, that the twins' mother used to make them smile, and how on earth she knew about the old desks in the shed. But by the time we rose to salute the flag and sing our morning song, I had it figured out. Father must have briefed her on the way home last night. The twins were an ever present concern of the whole Group and we were all especially anxious to have their first year a successful one. Also, Father knew the smile routine and where the old desks were stored. As for the spray of leaves, well, some did grow this low on the mountain and frost is tricky at leaf-turning time.

So school was launched and went along smoothly. Miss Carmody was a good teacher and even the Kroginolds found their studies interesting.

They hadn't tried any tricks since Jemmy threatened them. That is, except that silly deal with the chalk. Miss Carmody was explaining something on the board and was groping sideways for the chalk to add to the lesson. Jake was deliberately lifting the chalk every time she almost had it. I was just ready to do something about it when Miss Carmody snapped her fingers with annoyance and grasped the chalk firmly. Jake caught my eye about then and shrank about six inches in girth and height. I didn't tell Jemmy, but Jake's fear that I might kept him straight for a long time.

The twins were really blossoming. They laughed and played with the rest of the kids and Jerry even went off occasionally with the other boys at noontime, coming back as disheveled and wet as the others after a dam-building session in the creek.

Miss Carmody fitted so well into the community and was so well-liked by us kids that it began to look like we'd finally keep a teacher all year. Already she had withstood some of the shocks that had sent our other teachers screaming. For instance. . . .

The first time Susie got a robin redbreast sticker on her bookmark for reading a whole page—six lines—perfectly, she lifted all the way back to her seat, literally walking about four inches in the air. I held my breath until she sat down and was caressing the glossy sticker with one finger, then I sneaked a cautious look at Miss Carmody. She was sitting very erect, her hands clutching both ends of her desk as though in the act of rising, a look of incredulous surprise on her face. Then she relaxed, shook her head and smiled, and busied herself with some papers.

I let my breath out cautiously. The last teacher but two went into hysterics when one of the girls absent-mindedly lifted back to her seat because her sore foot hurt. I had hoped Miss Carmody was tougher—and apparently she was.

The same week, one noon hour, Jethro came pelting up to the schoolhouse where Valancy—that's her first name and I call her by it when we are alone, after all she's only four years older than I—was helping me with that gruesome Tests and

Measurements I was taking by extension from Teachers' College.

"Hey, Karen!" he yelled through the window. "Can you come out a minute?"

"Why?" I yelled back, annoyed at the interruption just when I was trying to figure what was normal about a normal grade curve.

"There's need," yelled Jethro.

I put down my book. "I'm sorry, Valancy. I'll go see what's eating him."

"Should I come too?" she asked. "If something's wrong—"

"It's probably just some silly thing," I said, edging out fast. When one of The People says "There's need," that means Group business.

"Adonday Veeah!" I muttered at Jethro as we rattled down the steep rocky path to the creek. "What are you trying to do? Get us all in trouble? What's the matter?"

"Look," said Jethro, and there were the boys standing around an alarmed but proud Jerry and above their heads, poised in the air over a half-built rock dam, was a huge boulder.

"Who lifted that?" I gasped.

"I did," volunteered Jerry, blushing crimson.

I turned on Jethro. "Well, why didn't you platt the twishers on it? You didn't have to come running—"

"On *that?*" Jethro squeaked. "You know very well we're not allowed to *lift* anything that big let alone platt it. Besides," shamefaced, "I can't remember that dern girl stuff."

"Oh, Jethro! You're so stupid sometimes!" I turned to Jerry. "How on earth did you ever lift anything that big?"

He squirmed. "I watched Daddy at the mine once."

"Does he let you lift at home?" I asked severely.

"I don't know." Jerry squashed mud with one shoe, hanging his head. "I never lifted anything before."

"Well, you know better. You kids aren't allowed to lift anything an Outsider your age can't handle alone. And not even that if you can't platt it afterwards."

"I know it," Jerry was still torn between embarrassment and pride.

"Well, remember it," I said. And taking a handful of sun, I platted the twishers and set the boulder back on the hillside where it belonged.

Platting does come easier to the girls—sunshine platting, that is. Of course only the Old Ones do the sun-and-rain one and only the very Oldest of them all would dare the moonlight-

and-dark, that can move mountains. But that was still no excuse for Jethro to forget and run the risk of having Valancy see what she mustn't see.

It wasn't until I was almost back to the schoolhouse that it dawned on me. Jerry had lifted! Kids his age usually lift play stuff almost from the time they walk. That doesn't need platting because it's just a matter of a few inches and a few seconds so gravity manages the return. But Jerry and Susie never had. They were finally beginning to catch up. Maybe it *was* just The Crossing that slowed them down—and maybe only the Clarinades. In my delight, *I* forgot and lifted to the school porch without benefit of the steps. But Valancy was putting up pictures on the high, old-fashioned moulding just below the ceiling, so no harm was done. She was flushed from her efforts and asked me to bring the step stool so she could finish them. I brought it and steadied it for her—and then nearly let her fall as I stared. How had she hung those first four pictures before I got there?"

The weather was unnaturally dry all Fall. We didn't mind it much because rain with an Outsider around is awfully messy. We have to let ourselves get wet. But when November came and went and Christmas was almost upon us, and there was practically no rain and no snow at all, we all began to get worried. The creek dropped to a trickle and then to scattered puddles and then went dry. Finally the Old Ones had to spend an evening at the Group Reservoir doing something about our dwindling water supply. They wanted to get rid of Valancy for the evening, just in case, so Jemmy volunteered to take her to Kerry to the show. I was still awake when they got home long after midnight. Since I began to develop the Gift, I have long periods of restlessness when it seems I have no apartness but am of every person in the Group. The training I should start soon will help me shut out the others except when I want them. The only thing is that we don't know who is to train me. Since Grandmother died there has been no Sorter in our Group and because of The Crossing we have no books or records to help.

Anyway, I was awake and leaning on my window sill in the darkness. They stopped on the porch—Jemmy is bunking at the mine during his stint there. I didn't have to guess or use a Gift to read the pantomime before me. I closed my eyes and my mind as their shadows merged. Under their strong emotion, I could have had free access to their minds, but I had been watching them all Fall. I knew in a special way what passed between them, and I knew that Valancy often went to bed in tears and that Jemmy spent too many lonely hours on the Crag that juts out over the canyon from high on Old Baldy, as though he were trying to make his heart as inaccessible to Outsiders as the Crag is. I knew what he felt, but oddly enough I had never been able to sort Valancy since that first night. There was something very un-Outsiderish and also very un-Groupish about her mind and I couldn't figure what.

I heard the front door open and close and Valancy's light steps fading down the hall and then I felt Jemmy calling me outside. I put my coat on over my robe and shivered down the hall. He was waiting by the porch steps, his face still and unhappy in the faint moonlight.

"She won't have me," he said flatly.

"Oh, Jemmy!" I cried. "You asked her—"

"Yes," he said. "She said no."

"I'm so sorry." I huddled down on the top step to cover my cold ankles. "But Jemmy—"

"Yes, I know!" He retorted savagely. "She's an Outsider. I have no business even to want her. Well, if she'd have me, I wouldn't hesitate a minute. This Purity-of-the-Group deal is—"

". . . is fine and right," I said softly, "as long as it doesn't touch you personally? But think for a minute, Jemmy. Would you be able to live a life as an Outsider? Just think of the million and one restraints that you would have to impose on yourself—and for the rest of your life, too, or lose her after all. Maybe it's better to accept *No* now than to try to build something and ruin it completely later. And if there should be children . . ." I paused. "*Could* there be children, Jemmy?"

I heard him draw a sharp breath.

"We don't know," I went on. "We haven't had the occasion to find out. Do you want Valancy to be part of the first experiment?"

Jemmy slapped his hat viciously down on his thigh, then he laughed.

"You have the Gift," he said, though I had never told him. "Have you any idea, sister mine, how little you will be liked when you become an Old One?"

"Grandmother was well-liked," I answered placidly. Then I cried, "Don't *you* set me apart, darn you, Jemmy. Isn't it enough to know that among a different people, *I* am different? Don't *you* desert me now!" I was almost in tears.

Jemmy dropped to the step beside me and thumped my

shoulder in his old way. "Pull up your socks, Karen. We have to do what we have to do. I was just taking my mad out on you. What a world." He sighed heavily.

I huddled deeper in my coat, cold of soul.

"But the other one is gone," I whispered. "The Home."

And we sat there sharing the poignant sorrow that is a constant undercurrent among The People, even those of us who never actually saw The Home. Father says it's because of a sort of racial memory.

"But she didn't say no because she doesn't love me," Jemmy went on at last. "She does love me. She told me so."

"Then why not?" Sister-wise I couldn't imagine anyone turning Jemmy down.

Jemmy laughed—a short, unhappy laugh. "Because she is different."

"She's different?"

"That's what she said, as though it was pulled out of her. 'I can't marry,' she said. 'I'm different!' That's pretty good, isn't it, coming from an Outsider!"

"She doesn't know we're The People," I said. "She must feel that she is different from everyone. I wonder why?"

"I don't know. There's something about her, though. A kind of shield or wall that keeps us apart. I've never met anything like it in an Outsider or in one of The People either. Sometimes it's like meshing with one of us and then *bang!* I smash the daylights out of me against that stone wall."

"Yes, I know," I said. "I've felt it, too."

We listened to the silent past-midnight world and then Jemmy stood.

"Well, g'night, Karen. Be seeing you."

I stood up, too. "Good night, Jemmy." I watched him start off in the late moonlight. He turned at the gate, his face hidden in the shadows.

"But I'm not giving up," he said quietly. "Valancy is my love."

The next day was hushed and warm—unnaturally so for December in our hills. There was a kind of ominous stillness among the trees, and, threading thinly against the milky sky, the thin smokes of little brush fires pointed out the dryness of the whole country. If you looked closely you could see piling behind Old Baldy an odd bank of clouds, so nearly the color of the sky that it was hardly discernible, but puffy and summer-thunderheady.

All of us were restless in school, the kids reacting to the

weather, Valancy pale and unhappy after last night. I was bruising my mind against the blank wall in hers, trying to find some way I could help her.

Finally the thousand and one little annoyances were climaxed by Jerry and Susie scuffling until Susie was pushed out of the desk onto an open box of wet water colors that Debra for heaven only knows what reason had left on the floor by her desk. Susie shrieked and Debra sputtered and Jerry started a high silly giggle of embarrassment and delight. Valancy, without looking, reached for something to rap for order with and knocked down the old cracked vase full of drooping wildflowers and three-day-old water. The vase broke and flooded her desk with the foul-smelling deluge, ruining the monthly report she had almost ready to send in to the County School Superintendent.

For a stricken moment there wasn't a sound in the room, then Valancy burst into half-hysterical laughter and the whole room rocked with her. We all rallied around doing what we could to clean up Susie and Valancy's desk and then Valancy declared a holiday and decided that it would be the perfect time to go up-canyon to the slopes of Baldy and gather what greenery we could find to decorate our school room for the holidays.

We all take our lunches to school, so we gathered them up and took along a square tarp the boys had brought to help build the dam in the creek. Now that the creek was dry, they couldn't use it and it'd come in handy to sit on at lunch time and would serve to carry our greenery home in, too, stretcher-fashion.

Released from the school room, we were all loud and jubilant and I nearly kinked my neck trying to keep all the kids in sight at once to nip in the bud any thoughtless lifting or other Group activity. The kids were all so wild, they might forget.

We went on up-canyon past the kids' dam and climbed the bare, dry waterfalls that stair-step up to the Mesa. On the Mesa, we spread the tarp and pooled our lunches to make it more picnicky. A sudden hush from across the tarp caught my attention. Debra, Rachel and Lizbeth were staring horrified at Susie's lunch. She was calmly dumping out a half dozen *koomatka* beside her sandwiches.

Koomatka are almost the only plants that lasted through The Crossing. I think four *koomatka* survived in someone's personal effects. They were planted and cared for as tenderly as babies and now every household in the Group has a *koomatka* plant growing in some quiet spot out of casual sight. Their fruit is eaten not so much for nourishment as Earth knows nourishment, but as a last remembrance of all other similar delights that died with The Home. We always save *Koomatka* for special occasions. Susie must have sneaked some out when her mother wasn't looking. And there they were—across the table from an Outsider!

Before I could snap them to me or say anything, Valancy turned, too, and caught sight of the softly glowing bluey-green pile. Her eyes widened and one hand went out. She started to say something and then she dropped her eyes quickly and drew her hand back. She clasped her hands tightly together and the girls, eyes intent on her, scrambled the *koomatka* back into the sack and Lizbeth silently comforted Susie who had just realized what she had done. She was on the verge of tears at having betrayed The People to an Outsider.

Just then 'Kiah and Derek rolled across the picnic table fighting over a cupcake. By the time we salvaged our lunch from under them and they had scraped the last of the chocolate frosting off their T-shirts, the *koomatka* incident seemed closed. And yet, as we lay back resting a little to settle our stomachs, staring up at the smothery low-hanging clouds that had grown from the milky morning sky, I suddenly found myself trying to decide about Valancy's look when she saw the fruit. Surely it couldn't have been recognition!

At the end of our brief siesta, we carefully buried the remains of our lunch—the hill was much too dry to think of burning it—and started on again. After a while, the slope got steeper and the stubborn tangle of manzanita tore at our clothes and scratched our legs and grabbed at the rolled-up tarp until we all looked longingly at the free air above it. If Valancy hadn't been with us we could have lifted over the worst and saved all this trouble. But we blew and panted for a while and then struggled on.

After an hour or so, we worked out onto a rocky knoll that leaned against the slope of Baldy and made a tiny island in the sea of manzanita. We all stretched out gratefully on the crumbling granite outcropping, listening to our heart-beats slowing.

Then Jethro sat up and sniffed. Valancy and I alerted. A sudden puff of wind from the little side canyon brought the acrid pungency of burning brush to us. Jethro scrambled along the narrow ridge to the slope of Baldy and worked his way around out of sight into the canyon. He came scrambling back, half lifting, half running.

"Awful!" he panted. "It's awful! The whole canyon ahead is on fire and it's coming this way fast!"

Valancy gathered us together with a glance.

"Why didn't we see the smoke?" she asked tensely. "There wasn't any smoke when we left the schoolhouse."

"Can't see this slope from school," he said. "Fire could burn over a dozen slopes and we'd hardly see the smoke. This side of Baldy is a rim fencing in an awful mess of canyons."

"What'll we do?" quavered Lizbeth, hugging Susie to her.

Another gust of wind and smoke set us all to coughing and through my streaming tears, I saw a long lapping tongue of fire reach around the canyon wall.

Valancy and I looked at each other. I couldn't sort her mind, but mine was a panic, beating itself against the fire and then against the terrible tangle of manzanita all around us. Bruising against the possibility of lifting out of danger, then against the fact that none of the kids was capable of sustained progressive self-lifting for more than a minute or so and how could we leave Valancy? I hid my face in my hands to shut out the acres and acres of tinder-dry manzanita that would blaze like a torch at the first touch of fire. If only it would rain! You can't *set* fire to wet manzanita, but after these long months of drought—!

I heard the younger children scream and looked up to see Valancy staring at me with an intensity that frightened me even as I saw fire standing bright and terrible behind her at the mouth of the canyon.

Jake, yelling hoarsely, broke from the group and lifted a yard or two over the manzanita before he tangled his feet and fell helpless into the ugly, angled branches.

"Get under the tarp!" Valancy's voice was a whip-lash. "All of you get under the tarp!"

"It won't do any good," bellowed 'Kiah. "It'll burn like paper!"

"Get—under—the—tarp!" Valancy's spaced, icy words drove us to unfolding the tarp and spreading it to creep under. I lifted (hoping even at this awful moment that Valancy wouldn't see me) over to Jake and yanked him back to his feet. I couldn't lift with him so I pushed and prodded and half-carried him back through the heavy surge of black smoke to the tarp and shoved him under. Valancy was standing, back to the fire, so changed and alien that I shut my eyes against her and started to crawl in with the other kids.

And then she began to speak. The rolling, terrible thunder of her voice shook my bones and I swallowed a scream. A surge of fear swept through our huddled group and shoved me back out from under the tarp.

Till I die, I'll never forget Valancy standing there tense and taller than life against the rolling convulsive clouds of smoke, both her hands outstretched, fingers wide apart as the measured terror of her voice went on and on in words that plague me because I should have known them and didn't. As I watched. I felt an icy cold gather, a paralyzing, unearthly cold that froze the tears on my tensely upturned face.

And then lightning leaped from finger to finger of her lifted hands. And lightning answered in the clouds above her. With a toss of her hands she threw the cold, the lightning, the sullen shifting smoke upward, and the roar of the racing fire was drowned in a hissing roar of down-drenching rain.

I knelt there in the deluge, looking for an eternal second into her drained, despairing, hopeless eyes before I caught her just in time to keep her head from banging on the granite as she pitched forward, inert.

Then as I sat there cradling her head in my lap, shaking with cold and fear, with the terrified wailing of the kids behind me, I heard Father shout and saw him and Jemmy and Darcy Clarinade in the old pick-up, lifting over the steaming streaming manzanita, over the trackless mountainside through the rain to us. Father lowered the truck until one of the wheels brushed a branch and spun lazily, then the three of them lifted all of us up to the dear familiarity of that beat-up old jalopy.

Jemmy received Valancy's limp body into his arms and crouched in back, huddling her in his arms, for the moment hostile to the whole world that had brought his love to such a pass.

We kids clung to Father in an ecstasy of relief. He hugged us all tight to him, then he raised my face.

"Why did it rain?" he asked sternly, every inch an Old One while the cold downpour dripped off the ends of my hair and he stood dry inside his Shield.

"I don't know," I sobbed, blinking my streaming eyes against his sternness. "Valancy did it . . . with lightning . . . it was cold . . . she talked. . . ." Then I broke down completely, plumping down on the rough floor boards and, in spite of my age, howling right along with the other kids.

It was a silent, solemn group that gathered in the schoolhouse that evening. I sat at my desk with my hands folded stiffly in front of me, half scared of my own People. This was the first official meeting of the Old Ones I'd ever attended. They all sat in desks, too, except the Oldest who sat in Valancy's chair. Valancy sat stony-faced in the twin's desk, but her nervous fingers shredded one kleenex after another as she waited.

The Oldest rapped the side of the desk with his cane and turned his sightless eyes from one to another of us.

"We're all here," he said, "to inquire——"

"Oh, stop it!" Valancy jumped up from her seat. "Can't you fire me without all this rigmarole? I'm used to it. Just say go and I'll go!" She stood trembling.

"Sit down, Miss Carmody," said the Oldest. And Valancy sat down meekly.

"Where were you born?" asked the Oldest quietly.

"What does it matter?" flared Valancy. Then resignedly, "It's in my application. Vista Mar, California."

"And your parents?"

"I don't know."

There was a stir in the room.

"Why not?"

"Oh, this is so unnecessary!" cried Valancy. "But if you *have* to know, both my parents were foundlings. They were found wandering in the streets after a big explosion and fire in Vista Mar. An old couple who lost everything in the fire took them in. When they grew up, they married. I was born. They died. Can I go now?"

A murmur swept the room.

"Why did you leave your other jobs?" asked Father.

Before Valancy could answer, the door was flung open and Jemmy stalked defiantly in.

"Go!" said the Oldest.

"Please," said Jemmy, deflating suddenly. "Let me stay. It concerns me too."

The Oldest fingered his cane and then nodded. Jemmy half-smiled with relief and sat down in a back seat.

"Go on," said the Oldest One to Valancy.

"All right then," said Valancy. "I lost my first job because I—well—I guess you'd call it levitated—to fix a broken blind in my room. It was stuck and I just . . . went up . . . in the air until I unstuck it. The principal saw me. He couldn't believe it and it scared him so he fired me." She paused expectantly.

The Old Ones looked at one another and my silly, confused mind began to add up columns that only my lack of common sense had kept from giving totals long ago.

"And the other one?" The Oldest leaned his cheek on his doubled-up hand as he bent forward.

Valancy was taken aback and she flushed in confusion.

"Well," she said hesitantly, "I called my books to me—I mean they were on my desk. . . ."

"We know what you mean," said The Oldest.
"You know!" Valancy looked dazed.
The Oldest stood up.
"Valancy Carmody, open your mind!"
Valancy stared at him and then burst into tears.
"I can't, I can't," she sobbed. "It's been too long. I can't let anyone in. I'm different. I'm alone. Can't you understand? They all died. I'm alien!"
"You are alien no longer," said the Oldest. "You are home now, Valancy." He motioned to me. "Karen, go in to her."
So I did. At first the wall was still there; then with a soundless cry, half anguish and half joy, the wall went down and I was with Valancy. I saw all the secrets that had cankered in her since her parents died—the parents who were of The People.
They had been reared by the old couple who were not only of The People but had been The Oldest of the whole Crossing.
I tasted with her the hidden frightening things—the need for living as an Outsider, the terrible need for concealing all her differences and suppressing all the extra Gifts of The People, the ever present fear of betraying herself and the awful lostness that came when she thought she was the last of The People.
And then suddenly *she* came in to *me* and my mind was flooded with a far greater presence than I had ever before experienced.
My eyes flew open and I saw all of the Old Ones staring at Valancy. Even the Oldest had his face turned to her, wonder written as widely on his scarred face as on the others.
He bowed his head and made The Sign. "The lost persuasions and designs," he murmured. "She has them all."
And then I knew that Valancy, Valancy who had wrapped herself so tightly against the world to which any thoughtless act might betray her that she had lived with us all this time without our knowing about her or she about us, was one of us. Not only one of us but such a one as had not been since Grandmother died—and even beyond that. My incoherent thoughts cleared to one.
Now I would have someone to train me. Now I could become a sorter—but only second to her.
I turned to share my wonder with Jemmy. He was looking at Valancy as The People must have looked at The Home in the last hour. Then he turned to the door.
Before I could draw a breath, Valancy was gone from me and from the Old Ones and Jemmy was turning to her outstretched hands.

Then I bolted for the outdoors and rushed like one possessed down the lane, lifting and running until I staggered up our porch steps and collapsed against Mother, who had heard me coming.

"Oh, Mother!" I cried. "She's one of us! She's Jemmy's love! She's wonderful!" And I burst into noisy sobs in the warm comfort of Mother's arms.

So now I don't have to go Outside to become a teacher. We have a permanent one. But I'm going anyway. I want to be as much like Valancy as I can and she has her degree. Besides I can use the discipline of living Outside for a year.

I have so much to learn and so much training to go through, but Valancy will always be there with me. I won't be set apart alone because of The Gift.

Maybe I shouldn't mention it, but one reason I want to hurry my training is that we're going to try to locate the other People. None of the boys here please me.

THE MOON IS GREEN

By Fritz Leiber

"Effie! What the devil are you up to?"

Her husband's voice, chopping through her mood of terrified rapture, made her heart jump like a startled cat, yet by some miracle of feminine self-control her body did not show a tremor.

Dear God, she thought, *he mustn't see it. It's so beautiful, and he always kills beauty.*

"I'm just looking at the Moon," she said listlessly. "It's green."

Mustn't, mustn't see it. And now, with luck, he wouldn't. For the face, as if it also heard and sensed the menace in the voice, was moving back from the window's glow into the outside dark, but slowly, reluctantly, and still faunlike, pleading, cajoling, tempting, and incredibly beautiful.

"Close the shutters at once, you little fool, and come away from the window!"

"Green as a beer bottle," she went on dreamily, "green as emeralds, green as leaves with sunshine striking through them and green grass to lie on." She couldn't help saying those last words. They were her token to the face, even though it couldn't hear.

"Effie!"

She knew what that last one meant. Wearily she swung shut the ponderous lead inner shutters and drove home the heavy bolts. That hurt her fingers; it always did, but he mustn't know that.

"You know that those shutters are not to be touched! Not for five more years at least!"

"I only wanted to look at the Moon," she said, turning around, and then it was all gone—the face, the night, the Moon, the magic—and she was back in the grubby, stale little hole, facing an angry, stale little man. It was then that the

Copyright, 1952, by Galaxy Publishing Corporation, and originally published in *Galaxy Science Fiction*, April, 1952.

eternal thud of the air-conditioning fans and the crackle of the electrostatic precipitators that sieved out the dust reached her consciousness again like the bite of a dentist's drill.

"Only wanted to look at the Moon!" he mimicked her in falsetto. "Only wanted to die like a little fool and make me that much more ashamed of you!" Then his voice went gruff and professional. "Here, count yourself."

She silently took the Geiger counter he held at arm's length, waited until it settled down to a steady ticking slower than a clock—due only to cosmic rays and indicating nothing dangerous—and then began to comb her body with the instrument. First her head and shoulders, then out along her arms and back along their under side. There was something oddly voluptuous about her movements, although her features were gray and sagging.

The ticking did not change its tempo until she came to her waist. Then it suddenly spurted, clicking faster and faster. Her husband gave an excited grunt, took a quick step forward, froze. She goggled for a moment in fear, then grinned foolishly, dug in the pocket of her grimy apron and guiltily pulled out a wrist watch.

He grabbed it as it dangled from her fingers, saw that it had a radium dial, cursed, heaved it up as if to smash it on the floor, but instead put it carefully on the table.

"You imbecile, you incredible imbecile," he softly chanted to himself through clenched teeth, with eyes half closed.

She shrugged faintly, put the Geiger counter on the table, and stood there slumped.

He waited until the chanting had soothed his anger before speaking again. He said quietly, "I do suppose you still realize the sort of world you're living in?"

She nodded slowly, staring at nothingness. Oh, she realized all right, realized only too well. It was the world that hadn't realized. The world that had gone on stockpiling hydrogen bombs. The world that had put those bombs in cobalt shells, although it had promised it wouldn't, because the cobalt made them much more terrible and cost no more. The world that had started throwing those bombs, always telling itself that it hadn't thrown enough of them yet to make the air really dangerous with the deadly radioactive dust that came from the cobalt. Thrown them and kept on throwing until the danger point, where air and ground would become fatal to all human life, was approached.

Then, for about a month, the two great enemy groups had

hesitated. And then each, unknown to the other, had decided it could risk one last gigantic and decisive attack without exceeding the danger point. It had been planned to strip off the cobalt cases, but someone forgot and then there wasn't time. Besides, the military scientists of each group were confident that the lands of the other had got the most dust. The two attacks came within an hour of each other.

After that, the Fury. The Fury of doomed men who think only of taking with them as many as possible of the enemy, and in this case—they hoped—all. The Fury of suicides who know they have botched up life for good. The Fury of cocksure men who realized they have been outsmarted by fate, the enemy, and themselves, and know that they will never be able to improvise a defense when arraigned before the high court of history—and whose unadmitted hope is that there will be no high court of history left to arraign them. More cobalt bombs were dropped during the Fury than in all the preceding years of the war.

After the Fury, the Terror. Men and women with death sifting into their bones through their nostrils and skin, fighting for bare survival under a dust-hazed sky that played fantastic tricks with the light of Sun and Moon, like the dust from Krakatoa that drifted around the world for years. Cities, countryside, and air were alike poisoned, alive with deadly radiation.

The only realistic chance for continued existence was to retire, for the five or ten years the radiation would remain deadly, to some well-sealed and radiation-shielded place that must also be copiously supplied with food, water, power, and a means of air-conditioning.

Such places were prepared by the far-seeing, seized by the stronger, defended by them in turn against the desperate hordes of the dying . . . until there were no more of those.

After that, only the waiting, the enduring. A mole's existence, without beauty or tenderness, but with fear and guilt as constant companions. Never to see the sun, to walk among the trees—or even know if there were still trees.

Oh, yes, she realized what the world was like.

"You understand, too, I suppose, that we were allowed to reclaim this ground-level apartment only because the Committee believed us to be responsible people, and because I've been making a damn good showing lately?"

"Yes, Hank."

"I thought you were eager for privacy. You want to go back to the basement tenements?"

God, no! Anything rather than that fetid huddling, that shameless communal sprawl. And yet, was this so much better? The nearness to the surface was meaningless; it only tantalized. And the privacy magnified Hank.

She shook her head dutifully and said, "No, Hank."

"Then why aren't you careful? I've told you a million times, Effie, that glass is no protection against the dust that's outside that window. The lead shutter must never be touched! If you make one single slip like that and it gets around, the Committee will send us back to the lower levels without blinking an eye. And they'll think twice before trusting me with any important jobs."

"I'm sorry, Hank."

"Sorry? What's the good of being sorry? The only thing that counts is never to make a slip! Why the devil do you do such things, Effie? What drives you to it?"

She swallowed. "It's just that it's so dreadful being cooped up like this," she said hesitatingly, "shut away from the sky and the sun. I'm just hungry for a little beauty."

"And do you suppose I'm not?" he demanded. "Don't you suppose I want to get outside, too, and be carefree and have a good time? But I'm not so damn selfish about it. I want my children to enjoy the sun, and my children's children. Don't you see that that's the all-important thing and that we have to behave like mature adults and make sacrifices for it?"

"Yes, Hank."

He surveyed her slumped figure, her lined and listless face. "You're a fine one to talk about hunger for beauty," he told her. Then his voice grew softer, more deliberate. "You haven't forgotten, have you, Effie, that until last month the Committee was so concerned about your sterility? That they were about to enter my name on the list of those waiting to be allotted a free woman? Very high on the list, too!"

She could nod even at that one, but not while looking at him. She turned away. She knew very well that the Committee was justified in worrying about the birth rate. When the community finally moved back to the surface again, each additional healthy young person would be an asset, not only in the struggle for bare survival, but in the resumed war against Communism which some of the Committee members still counted on.

It was natural that they should view a sterile woman with disfavor, and not only because of the waste of her husband's germ-plasm, but because sterility might indicate that she had suffered more than the average from radiation. In that case, if she did bear children later on, they would be more apt to carry a defective heredity, producing an undue number of monsters and freaks in future generations, and so contaminating the race.

Of course she understood it. She could hardly remember the time when she didn't. Years ago? Centuries? There wasn't much difference in a place where time was endless.

His lecture finished, her husband smiled and grew almost cheerful.

"Now that you're going to have a child, that's all in the background again. Do you know, Effie, that when I first came in, I had some very good news for you? I'm to become a member of the Junior Committee and the announcement will be made at the banquet tonight." He cut short her mumbled congratulations. "So brighten yourself up and put on your best dress. I want the other Juniors to see what a handsome wife the new member has got." He paused. "Well, get a move on!"

She spoke with difficulty, still not looking at him. "I'm terribly sorry, Hank, but you'll have to go alone. I'm not well."

He straightened up with an indignant jerk. "There you go again! First that infantile, inexcusable business of the shutters, and now this! No feeling for my reputation at all. Don't be ridiculous, Effie. You're coming!"

"Terribly sorry," she repeated blindly, "but I really can't. I'd just be sick. I wouldn't make you proud of me at all."

"Of course you won't," he retorted sharply. "As it is, I have to spend half my energy running around making excuses for you—why you're so odd, why you always seem to be ailing, why you're always stupid and snobbish and say the wrong thing. But tonight's really important, Effie. It will cause a lot of bad comment if the new member's wife isn't present. You know how just a hint of sickness starts the old radiation-disease rumor going. You've *got* to come, Effie."

She shook her head helplessly.

"Oh, for heaven's sake, come on!" he shouted, advancing on her. "This is just a silly mood. As soon as you get going, you'll snap out of it. There's nothing really wrong with you at all."

He put his hand on her shoulder to touch her around, and at his touch her face suddenly grew so desperate and gray that for a moment he was alarmed in spite of himself.

"Really?" he asked, almost with a note of concern.

She nodded miserably.

"Hmm!" He stepped back and strode about irresolutely. "Well, of course, if that's the way it is . . ." He checked himself and a sad smile crossed his face. "So you don't care enough about your old husband's success to make one supreme effort in spite of feeling bad?"

Again the helpless headshake. "I just can't go out tonight, under any circumstances." And her gaze stole toward the lead shutters.

He was about to say something when he caught the direction of her gaze. His eyebrows jumped. For seconds he stared at her incredulously, as if some completely new and almost unbelievable possibility had popped into his mind. The look of incredulity slowly faded, to be replaced by a harder, more calculating expression. But when he spoke again, his voice was shockingly bright and kind.

"Well, it can't be helped, naturally, and I certainly wouldn't want you to go if you weren't able to enjoy it. So you hop right into bed and get a good rest. I'll run over to the men's dorm to freshen up. No, really, I don't want you to have to make any effort at all. Incidentally, Jim Barnes isn't going to be able to come to the banquet either—touch of the old 'flu, he tells me, of all things."

He watched her closely as he mentioned the other man's name, but she didn't react noticeably. In fact, she hardly seemed to be hearing his chatter.

"I got a bit sharp with you, I'm afraid, Effie," he continued contritely. "I'm sorry about that. I was excited about my new job and I guess that was why things upset me. Made me feel let down when I found you weren't feeling as good as I was. Selfish of me. Now you get into bed right away and get well. Don't worry about me a bit. I know you'd come if you possibly could. And I know you'll be thinking about me. Well, I must be off now."

He started toward her, as if to embrace her, then seemed to think better of it. He turned back at the doorway and said, emphasizing the words, "You'll be completely alone for the next four hours." He waited for her nod, then bounced out.

She stood still until his footsteps died away. Then she straightened up, walked over to where he'd put down the wristwatch, picked it up and smashed it hard on the floor. The crystal shattered, the case flew apart, and something went *zing!*

She stood there breathing heavily. Slowly her sagged features lifted, formed themselves into the beginning of a smile. She stole another look at the shutters. The smile became more definite. She felt her hair, wet her fingers and ran them along her hairline and back over her ears. After wiping her hands on her apron, she took it off. She straightened her dress, lifted her head with a little flourish, and stepped smartly toward the window.

Then her face went miserable again and her steps slowed.

No, it couldn't be, and it won't be, she told herself. It had been just an illusion, a silly romantic dream that she had somehow projected out of her beauty-starved mind and given a moment's false reality. There couldn't be anything alive outside. There hadn't been for two whole years.

And if there conceivably were, it would be something altogether horrible. She remembered some of the pariahs—hairless, witless creatures, with radiation welts crawling over their bodies like worms, who had come begging for succor during the last months of the Terror—and been shot down. How they must have hated the people in refuges!

But even as she was thinking these things, her fingers were caressing the bolts, gingerly drawing them, and she was opening the shutters gently, apprehensively.

No, there couldn't be anything outside, she assured herself wryly, peering out into the green night. Even her fears had been groundless.

But the face came floating up toward the window. She started back in terror, then checked herself.

For the face wasn't horrible at all, only very thin, with full lips and large eyes and a thin proud nose like the jutting beak of a bird. And no radiation welts or scars marred the skin, olive in the tempered moonlight. It looked, in fact, just as it had when she had seen it for the first time.

For a long moment the face stared deep, deep into her brain. Then the full lips smiled and a half-clenched, thin-fingered hand materialized itself from the green darkness and rapped twice on the grimy pane.

Her heart pounding, she furiously worked the little crank that opened the window. It came unstuck from the frame with a tiny explosion of dust and a *zing* like that of the watch, only louder. A moment later it swung open wide and a puff of incredibly fresh air caressed her face and the inside of her nostrils, stinging her eyes with unanticipated tears.

The man outside balanced on the sill, crouching like a faun, head high, one elbow on knee. He was dressed in scarred, snug trousers and an old sweater.

"Is it tears I get for a welcome?" he mocked her gently in a musical voice. "Or are those only to greet God's own breath, the air?"

He swung down inside and now she could see he was tall. Turning, he snapped his fingers and called, "Come, puss."

A black cat with a twisted stump of a tail and feet like small boxing gloves and ears almost as big as rabbits' hopped clumsily in view. He lifted it down, gave it a pat. Then, nodding familiarly to Effie, he unstrapped a little pack from his back and laid it on the table.

She couldn't move. She even found it hard to breathe.

"The window," she finally managed to get out.

He looked at her inquiringly, caught the direction of her stabbing finger. Moving without haste, he went over and closed it carelessly.

"The shutters, too," she told him, but he ignored that, looking around.

"It's a snug enough place you and your man have," he commented. "Or is it that this is a free-love town or a harem spot, or just a military post?" He checked her before she could answer. "But let's not be talking about such things now. Soon enough I'll be scared to death for both of us. Best enjoy the kick of meeting, which is always good for twenty minutes at the least." He smiled at her rather shyly. "Have you food? Good, then bring it."

She set cold meat and some precious canned bread before him and had water heating for coffee. Before he fell to, she shredded a chunk of meat and put it on the floor for the cat, which left off its sniffing inspection of the walls and ran up eagerly mewing. Then the man began to eat, chewing each mouthful slowly and appreciatively.

From across the table Effie watched him, drinking in his every deft movement, his every cryptic quirk of expression. She attended to making the coffee, but that took only a moment. Finally she could contain herself no longer.

"What's it like up there?" she asked breathlessly. "Outside, I mean."

He looked at her oddly for quite a space. Finally, he said flatly, "Oh, it's a wonderland for sure, more amazing than you tombed folk could ever imagine. A veritable fairyland." And he quickly went on eating.

"No, but really," she pressed.

Noting her eagerness, he smiled and his eyes filled with playful tenderness. "I mean it, on my oath," he assured her. "You think the bombs and the dust made only death and ugliness. That was true at first. But then, just as the doctors foretold, they changed the life in the seeds and loins that were brave enough to stay. Wonders bloomed and walked." He broke off suddenly and asked, "Do any of you ever venture outside?"

The Moon Is Green

"A few of the men are allowed to," she told him, "for short trips in special protective suits, to hunt for canned food and fuels and batteries and things like that."

"Aye, and those blind-souled slugs would never see anything but what they're looking for," he said, nodding bitterly. "They'd never see the garden where a dozen buds blossom where one did before, and the flowers have petals a yard across, with stingles bees big as sparrows gently supping their nectar. Housecats grown spotted and huge as leopards (not little runts like Joe Louis here) stalk through those gardens. But they're gentle beasts, no more harmful than the rainbow-scaled snakes that glide around their paws, for the dust burned all the murder out of them, as it burned itself out.

"I've even made up a little poem about that. It starts, 'Fire can hurt me, or water, or the weight of Earth. But the dust is my friend.' Oh, yes, and then the robins like cockatoos and squirrels like a princess's ermine! All under a treasure chest of Sun and Moon and stars that the dust's magic powder changes from ruby to emerald and sapphire and amethyst and back again. Oh, and then the new children—"

"You're telling the truth?" she interrupted him, her eyes brimming with tears. "You're not making it up?"

"I am not," he assured her solemnly. "And if you could catch a glimpse of one of the new children, you'd never doubt me again. They have long limbs as brown as this coffee would be if it had lots of fresh cream in it, and smiling delicate faces and the whitest teeth and the finest hair. They're so nimble that I—a sprightly man and somewhat enlivened by the dust—feel like a cripple beside them. And their thoughts dance like flames and make me feel a very imbecile.

"Of course, they have seven fingers on each hand and eight toes on each foot, but they're more beautiful for that. They have large pointed ears that the sun shines through. They play in the garden, all day long, slipping among the great leaves and blooms, but they're so swift that you can hardly see them, unless one chooses to stand still and look at you. For that matter, you have to look a bit hard for all these things I'm telling you."

"But it is true?" she pleaded.

"Every word of it," he said, looking straight into her eyes. He put down his knife and fork. "What's your name?" he asked softly. "Mine's Patrick."

"Effie," she told him.

He shook his head. "That can't be," he said. Then his face brightened. "Euphemia," he exclaimed. "That's what Effie is short for. Your name is Euphemia." As he said that, looking at her, she suddenly felt beautiful. He got up and came around the table and stretched out his hand toward her.

"Euphemia—" he began.

"Yes?" she answered huskily, shrinking from him a little, but looking up sideways, and very flushed.

"Don't either of you move," Hank said.

The voice was flat and nasal because Hank was wearing a nose respirator that was just long enough to suggest an elephant's trunk. In his right hand was a large blue-black automatic pistol.

They turned their faces to him. Patrick's was abruptly alert, shifty. But Effie's was still smiling tenderly, as if Hank could not break the spell of the magic garden and should be pitied for not knowing about it.

"You little—" Hank began with an almost gleeful fury, calling her several shameful names. He spoke in short phrases, closing tight his unmasked mouth between them while he sucked in breath through the respirator. His voice rose in a crescendo. "And not with a man of the community, but a pariah! A *pariah!*"

"I hardly know what you're thinking, man, but you're quite wrong," Patrick took the opportunity to put in hurriedly, conciliatingly. "I just happened to be coming by hungry tonight, a lonely tramp, and knocked at the window. Your wife was a bit foolish and let kindheartedness get the better of prudence—"

"Don't think you've pulled the wool over my eyes, Effie," Hank went on with a screechy laugh, disregarding the other man completely. "Don't think I don't know why you're suddenly going to have a child after four long years."

At that moment the cat came nosing up to his feet. Patrick watched him narrowly, shifting his weight forward a little, but Hank only kicked the animal aside without taking his eyes off them.

"Even that business of carrying the wristwatch in your pocket instead of on your arm," he went on with channeled hysteria. "A neat bit of camouflage, Effie. Very neat. And telling me it was my child, when all the while you've been seeing him for months!"

"Man, you're mad; I've not touched her!" Patrick denied hotly though still calculating, and risked a step forward, stopping when the gun instantly swung his way.

"Pretending you were going to give me a healthy child,"

Hank raved on, "when all the while you knew it would be—either in body or germ plasm—a thing like *that!*"

He waved his gun at the malformed cat, which had leaped to the top of the table and was eating the remains of Patrick's food, though its watchful green eyes were fixed on Hank.

"I should shoot him down!" Hank yelled, between sobbing, chest-racking inhalations through the mask. "I should kill him this instant for the contaminated pariah he is!"

All this while Effie had not ceased to smile compassionately. Now she stood up without haste and went to Patrick's side. Disregarding his warning, apprehensive glance, she put her arm lightly around him and faced her husband.

"Then you'd be killing the bringer of the best news we've ever had," she said, and her voice was like a flood of some warm sweet liquor in that musty, hate-charged room. "Oh, Hank, forget your silly, wrong jealousy and listen to me. Patrick here has something wonderful to tell us."

Hank stared at her. For once he screamed no reply. It was obvious that he was seeing for the first time how beautiful she had become, and that the realization jolted him terribly.

"What do you mean?" he finally asked unevenly, almost fearfully.

"I mean that we no longer need to fear the dust," she said, and now her smile was radiant. "It never really did hurt people the way the doctors said it would. Remember how it was with me, Hank, the exposure I had and recovered from, although the doctors said I wouldn't at first—and without even losing my hair? Hank, those who were brave enough to stay outside, and who weren't killed by terror and suggestion and panic—they adapted to the dust. They changed, but they changed for the better. Everything—"

"Effie, he told you lies!" Hank interrupted, but still in that same agitated, broken voice, cowed by her beauty.

"Everything that grew or moved was purified," she went on ringingly." You men going outside have never seen it, because you've never had eyes for it. You've been blinded to beauty, to life itself. And now all the power in the dust has gone and faded, anyway, burned itself out. That's true, isn't it?"

She smiled at Patrick for confirmation. His face was strangely veiled, as if he were calculating obscure changes. He might have given a little nod; at any rate, Effie assumed that he did, for she turned back to her husband.

"You see, Hank? We can all go out now. We need never fear the dust again. Patrick is a living proof of that," she continued triumphantly, standing straighter, holding him a little tighter. "Look at him. Not a scar or a sign, and he's been out in the dust for years. How could he be this way, if the dust hurt the brave? Oh, believe me, Hank! Believe what you see. Test it if you want. Test Patrick here."

"Effie, you're all mixed up. You don't know—" Hank faltered, but without conviction of any sort.

"Just test him," Effie repeated with utter confidence, ignoring—not even noticing—Patrick's warning nudge.

"All right," Hank mumbled. He looked at the stranger dully. "Can you count?" he asked.

Patrick's face was a complete enigma. Then he suddenly spoke, and his voice was like a fencer's foil—light, bright, alert, constantly playing, yet utterly on guard.

"Can I count? Do you take me for a complete simpleton, man? Of course I can count!"

"Then count yourself," Hank said, barely indicating the table.

"Count myself, should I?" the other retorted with a quick facetious laugh. "Is this a kindergarten? But if you want me to, I'm willing." His voice was rapid. "I've two arms, and two legs, that's four. And ten fingers and ten toes—you'll take my word for them?—that's twenty-four. A head, twenty-five. And two eyes and a nose and a mouth—"

"With this, I mean," Hank said heavily, advanced to the table, picked up the Geiger counter, switched it on, and handed it across the table to the other man.

But while it was still an arm's length from Patrick, the clicks began to mount furiously, until they were like the chatter of a pigmy machine gun. Abruptly the clicks slowed, but that was only the counter shifting to a new scaling circuit, in which each click stood for 512 of the old ones.

With those horrid, rattling little volleys, fear cascaded into the room and filled it, smashing like so much colored glass all the bright barriers of words Effie had raised against it. For no dreams can stand against the Geiger counter, the Twentieth Century's mouthpiece of ultimate truth. It was as if the dust and all the terrors of the dust had incarnated themselves in one dread invading shape that said in words stronger than audible speech, "Those were illusions, whistles in the dark. This is reality, the dreary, pitiless reality of the Burrowing Years."

Hank scuttled back to the wall. Through chattering teeth he babbled, ". . . enough radioactives . . . kill a thousand men . . . freak . . . a freak . . ." In his agitation he forgot for a moment to inhale through the respirator.

Even Effie—taken off guard, all the fears that had been drilled into her twanging like piano wires—shrank from the skeletal-seeming shape beside her, held herself to it only by desperation.

Patrick did it for her. He disengaged her arm and stepped briskly away. Then he whirled on them, smiling sardonically, and started to speak, but instead looked with distaste at the chattering Geiger counter he held between fingers and thumb.

"Have we listened to this racket long enough?" he asked.

Without waiting for an answer, he put down the instrument on the table. The cat hurried over to it curiously and the clicks began again to mount in a minor crescendo. Effie lunged for it frantically, switched it off, darted back.

"That's right," Patrick said with another chilling smile. "You do well to cringe, for I'm death itself. Even in death I could kill you, like a snake." And with that his voice took on the tones of a circus barker. "Yes, I'm a freak, as the gentleman so wisely said. That's what one doctor who dared talk with me for a minute told me before he kicked me out. He couldn't tell me why, but somehow the dust doesn't kill me. Because I'm a freak, you see, just like the men who ate nails and walked on fire and ate arsenic and stuck themselves through with pins. Step right up, ladies and gentlemen—only not too close!—and examine the man the dust can't harm. Rappaccini's child, brought up to date; his embrace, death!

"And now," he said, breathing heavily, "I'll get out and leave you in your damned lead cave."

He started toward the window. Hank's gun followed him shakingly.

"Wait!" Effie called in an agonized voice. He obeyed. She continued falteringly, "When we were together earlier, you didn't act as if . . ."

"When we were together earlier, I wanted what I wanted," he snarled at her. "You don't suppose I'm a bloody saint, do you?"

"And all the beautiful things you told me?"

"That," he said cruelly, "is just a line I've found that women fall for. They're all so bored and so starved for beauty—as *they* generally put it."

"Even the garden?" Her question was barely audible through the sobs that threatened to suffocate her.

He looked at her and perhaps his expression softened just a trifle.

"What's outside," he said flatly, "is just a little worse than either of you can imagine." He tapped his temple. "The garden's all here."

"You've killed it," she wept. "You've killed it in me. You've both killed everything that's beautiful. But you're worse," she screamed at Patrick, "because he only killed beauty once, but you brought it to life just so you could kill it again. Oh, I can't stand it! I won't stand it!" And she began to scream.

Patrick started toward her, but she broke off and whirled away from him to the window, her eyes crazy.

"You've been lying to us," she cried. "The garden's there. I know it is. But you don't want to share it with anyone."

"No, no, Euphemia," Patrick protested anxiously. "It's hell out there, believe me. I wouldn't lie to you about it."

"Wouldn't lie to me!" she mocked. "Are you afraid, too?"

With a sudden pull, she jerked open the window and stood before the blank green-tinged oblong of darkness that seemed to press into the room like a menacing, heavy, wind-urged curtain.

At that Hank cried out a shocked, pleading "Effie!"

She ignored him. "I can't be cooped up here any longer," she said. "And I won't, now that I know. I'm going to the garden."

Both men sprang at her, but they were too late. She leaped lightly to the sill, and by the time they had flung themselves against it, her footsteps were already hurrying off into the darkness.

"Effie, come back! Come back!" Hank shouted after her desperately, no longer thinking to cringe from the man beside him, or how the gun was pointed. "I love you, Effie. Come back!"

Patrick added his voice. "Come back, Euphemia. You'll be safe if you come back right away. Come back to your home."

No answer to that at all.

They both strained their eyes through the greenish murk. They could barely make out a shadowy figure about half a block down the near-black canyon of the dismal, dust-blown street, into which the greenish moonlight hardly reached. It seemed to them that the figure was scooping something up from the pavement and letting it sift down along its arms and over its bosom.

"Go out and get her, man," Patrick urged the other. "For if I go out for her, I warn you I won't bring her back. She said something about having stood the dust better than most, and that's enough for me."

But Hank, chained by his painfully learned habits and by something else, could not move.

And then a ghostly voice came whispering down the street, chanting, "Fire can hurt me, or water, or the weight of the Earth. But the dust is my friend."

Patrick spared the other man one more look. Then, without a word, he vaulted up and ran off.

Hank stood there. After perhaps a half minute he remembered to close his mouth when he inhaled. Finally he was sure the street was empty. As he started to close the window, there was a little *mew*.

He picked up the cat and gently put it outside. Then he did close the window, and the shutters, and bolted them, and took up the Geiger counter, and mechanically began to count himself.

SURVIVAL

By John Wyndham

As THE SPACEPORT bus trundled unhurriedly over the mile or more of open field that separated the terminal buildings from the embarkation hoist, Mrs. Feltham stared intently forward across the receding row of shoulders in front of her. The ship stood up on the plain like an isolated silver spire. Near its bow she could see the intense blue light which proclaimed it all but ready to take off. Among and around the great tailfins dwarf vehicles and little dots of men moved in a fuss of final preparations. Mrs. Feltham glared at the scene, at this moment loathing it, and all the inventions of men, with a hard, hopeless hatred.

Presently she withdrew her gaze from the distance and focussed it on the back of her son-in-law's head, a yard in front of her. She hated him, too.

She turned, darting a swift glance at the face of her daughter in the seat beside her. Alice looked pale; her lips were firmly set; her eyes fixed straight ahead.

Mrs. Feltham hesitated. Her glance returned to the spaceship. She decided on one last effort. Under cover of the bus noise she said:

"Alice, darling, it's not too late, even now, you know."

The girl did not look at her. There was no sign that she had heard, save that her lips compressed a little more firmly. Then they parted.

"Mother, please!" she said.

But Mrs. Feltham, once started, had to go on.

"It's for your own sake, darling. All you have to do is to say you've changed your mind."

The girl held a protesting silence.

"Nobody would blame you," Mrs. Feltham persisted. "They'd not think a bit the worse of you. After all, everybody knows that Mars is no place for—"

Copyright, 1951, by Standard Magazines, Inc., and originally published in *Thrilling Wonder Stories*, February, 1952.

Survival

"Mother, please stop it," interrupted the girl. The sharpness of her tone took Mrs. Feltham aback for a moment. She hesitated. But time was growing too short to allow herself the luxury of offended dignity. She went on:

"You're not used to the sort of life you'll have to live there, darling. Absolutely primitive. No kind of life for any woman. After all, dear, it is only a five-year appointment for David. I'm sure if he really loves you he'd rather know that you *are* safe here and waiting—"

The girl said, harshly:

"We've been over all this before, Mother. I tell you it's no good. I'm not a child. I've thought it out, and I've made up my mind."

Mrs. Feltham sat silent for some moments. The bus swayed on across the field, and the rocketship seemed to tower further into the sky.

"If you had a child of your own—" she said, half to herself. "—Well, I expect someday you will. Then you will begin to understand. . . ."

"I think it's you who don't understand," Alice said. "This is hard enough, anyway. You're only making it harder for me."

"My darling, I love you. I gave birth to you. I've watched over you always and I *know* you. I *know* this can't be the kind of life for you. If you were a hard, hoydenish kind of girl, well, perhaps—but you aren't, darling. You know quite well you aren't."

"Perhaps you don't know me quite as well as you imagine you do, Mother."

Mrs. Feltham shook her head. She kept her eyes averted, boring jealously into the back of her son-in-law's head.

"He's taken you right away from me," she said dully.

"That's not true, Mother. It's—well, I'm no longer a child. I'm a woman with a life of my own to live."

" 'Whither thou goest, I will go . . .' " said Mrs. Feltham reflectively. "But that doesn't really hold now, you know. It was all right for a tribe of nomads, but nowadays the wives of soldiers, sailors, pilots, spacemen—"

"It's more than that, Mother. You don't understand. I must become adult and real to myself. . . ."

The bus rolled to a stop, puny and toylike beside the ship that seemed too large ever to lift. The passengers got out and stood staring upwards along the shining side. Mr. Feltham put his arms round his daughter. Alice clung to him, tears in her eyes. In an unsteady voice he murmured:

"Goodbye, my dear. And all the luck there is."

He released her, and shook hands with his son-in-law.

"Keep her safe, David. She's everything—"

"I know. I will. Don't you worry."

Mrs. Feltham kissed her daughter farewell, and forced herself to shake hands with her son-in-law.

A voice from the hoist called: "All passengers aboard, please!"

The doors of the hoist closed. Mr. Feltham avoided his wife's eyes. He put his arm round her waist, and led her back to the bus in silence.

As they made their way, in company with a dozen other vehicles, back to the shelter of the terminal, Mrs. Feltham alternately dabbed her eyes with a wisp of white handkerchief and cast glances back at the spaceship standing tall, inert, and apparently deserted now. Her hand slid into her husband's.

"I can't believe it even now," she said. "It's so utterly unlike her. Would you ever have thought that our little Alice . . . ? Oh, why did she have to marry him . . .?" Her voice trailed to a whimper.

Her husband pressed her fingers, without speaking.

"It wouldn't be so surprising with some girls," she went on. "But Alice was always so quiet. I used to worry because she was so quiet—I mean in case she might become one of those timid bores. Do you remember how the other children used to call her Mouse?

"And now this! Five years in that dreadful place! Oh, she'll never stand it, Henry. I know she won't, she's not the type. Why didn't you put your foot down, Henry? They'd have listened to you. You could have stopped it."

Her husband sighed. "There are times when one can give advice, Miriam, though it's scarcely ever popular, but what one must not do is try to live other people's lives for them. Alice is a woman now, with her own rights. Who am I to say what's best for her?"

"But you could have stopped her going."

"Perhaps—but I didn't care for the price."

She was silent for some seconds, then her fingers tightened on his hand.

"Henry—Henry, I don't think we shall ever see them again. I feel it."

"Come, come, dear. They'll be back safe and sound, you'll see."

"You don't really believe that, Henry. You're just trying to cheer me up. Oh, why, why must she go to that horrible place?

She's so young. She could have waited five years. Why is she so stubborn, so hard—not like my little Mouse, at all?"

Her husband patted her hand reassuringly.

"You must try to stop thinking of her as a child, Miriam. She's not; she's a woman now and if all our women were mice, it would be a poor outlook for our survival. . . ."

The Navigating Officer of the s/r *Falcon* approached his Captain.

"The deviation, sir."

Captain Winters took the piece of paper held out to him.

"One point three six five degrees," he read out. "H'm. Not bad. Not at all bad, considering. South-east sector again. Why are nearly all deviations in the S.E. sector, I wonder, Mr. Carter?"

"Maybe they'll find out when we've been at the game a bit longer, sir. Right now it's just one of those things."

"Odd, all the same. Well, we'd better correct it before it gets any bigger."

The Captain loosened the expanding book-rack in front of him and pulled out a set of tables. He consulted them and scribbled down the result.

"Check, Mr. Carter."

The navigator compared the figures with the table, and approved.

"Good. How's she lying?" asked the Captain.

"Almost broadside, with a very slow roll, sir."

"You can handle it. I'll observe visually. Align her and stabilize. Ten seconds on starboard laterals at force two. She should take about thirty minutes, twenty seconds to swing over, but we'll watch that. Then neutralize with the port laterals at force two. Okay?"

"Very good, sir." The Navigating Officer sat down in the control chair, and fastened the belt. He looked over the keys and switches carefully.

"I'd better warn 'em. May be a bit of a jolt," said the Captain. He switched on the address system, and pulled the microphone bracket to him.

"Attention all! Attention all! We are about to correct course. There will be several impulses. None of them will be violent, but all fragile objects should be secured, and you are advised to seat yourselves and use the safety belts. The operation will take approximately half an hour and will start in five minutes from now. I shall inform you when it has been completed. That is all." He switched off.

"Some fool always thinks the ship's been holed by a meteor if you don't spoon it out," he added. "Have that woman in hysterics, most likely. Doesn't do any good." He pondered idly. "I wonder what the devil she thinks she's doing out here, anyway. A quiet little thing like that; what she ought to be doing is sitting in some village back home, knitting."

"She knits here," observed the Navigating Officer.

"I know—and think what it implies! What's the idea of that kind going to Mars? She'll be as homesick as hell, and hate every foot of the place on sight. That husband of hers ought to have had more sense. Comes damn near cruelty to children."

"It mightn't be his fault, sir. I mean, some of those quiet ones can be amazingly stubborn."

The captain eyed his officer speculatively.

"Well, I'm not a man of wide experience, but I know what I'd say to my wife if she thought of coming along."

"But you can't have a proper ding-dong with those quiet ones, sir. They kind of featherbed the whole thing, and then get their own way in the end."

"I'll overlook the implication of the first part of that remark, Mr. Carter, but out of this extensive knowledge of women can you suggest to me why the devil she is here if he didn't drag her along? It isn't as if Mars were domestically hazardous, like a convention."

"Well, sir—she strikes me as the devoted type. Scared of her own shadow ordinarily, but with an awful amount of determination when the right string's pulled. It's sort of—well, you've heard of ewes facing lions in defense of their cubs, haven't you?"

"Assuming that you mean lambs," said the Captain, "the answers would be, A: I've always doubted it; and, B: she doesn't have any."

"I was just trying to indicate the type, sir."

The Captain scratched his cheek with his forefinger.

"You may be right, but I know if I were going to take a wife to Mars, which heaven forbid, I'd feel a tough, gun-toting Momma was less of a liability. What's his job there?"

"Taking charge of a mining company office, I think."

"Office hours, huh? Well, maybe it'll work out someway, but I still say the poor little thing ought to be in her own kitchen. She'll spend half the time scared to death, and the rest of it pining for home comforts." He glanced at the clock. "They've had enough time to batten down the chamber-pots now. Let's get busy."

Survival

He fastened his own safety-belt, swung the screen in front of him on its pivot, switching it on as he did so, and leaned back watching the panorama of stars move slowly across it.

"All set, Mr. Carter?"

The Navigating Officer switched on a fuel line, and poised his right hand above a key.

"All set, sir."

"Okay. Straighten her up."

The Navigating Officer glued his attention to the pointers before him. He tapped the key beneath his fingers experimentally. Nothing happened. A slight double furrow appeared between his brows. He tapped again. Still there was no response.

"Get on with it, man," said the Captain irritably.

The Navigating Officer decided to try twisting her the other way. He tapped one of the keys under his left hand. This time there was response without delay. The whole ship jumped violently sideways and trembled. A crash jangled back and forth through the metal members around them like a diminishing echo.

Only the safety belt kept the Navigating Officer in his seat. He stared stupidly at the gyrating pointers before him. On the screen the stars were streaking across like a shower of fireworks. The Captain watched the display in ominous silence for a moment, then he said coldly:

"Perhaps when you have had your fun, Mr. Carter, you will kindly straighten her up."

The navigator pulled himself together. He chose a key, and pressed it. Nothing happened. He tried another. Still the needles on the dials revolved smoothly. A slight sweat broke out on his forehead. He switched to another fuel line, and tried again.

The Captain lay back in his chair, watching the heavens stream across his screen.

"Well?" he demanded, curtly.

"There's—no response, sir."

Captain Winters unfastened his safety-belt and clacked across the floor on his magnetic soles. He jerked his head for the other to get out of his seat, and took his place. He checked the fuel line switches. He pressed a key. There was no impulse: the pointers continued to turn without a check. He tried other keys, fruitlessly. He looked up and met the navigator's eyes. After a long moment he moved back to his own desk, and flipped a switch. A voice broke into the room:

"—would I know? All I know is that the old can's just bowling along head over elbow, and that ain't no kind of a way to run a bloody spaceship. If you ask me—"

"Jevons," snapped the Captain.

The voice broke off abruptly.

"Yes, sir?" it said, in a different tone.

"The laterals aren't firing."

"No, sir," the voice agreed.

"Wake up, man. I mean they *won't* fire. They're packed up."

"What—all of 'em, sir?"

"The only ones that have responded are the port laterals—and they shouldn't have kicked the way they did. Better send someone outside to look at 'em. I didn't like that kick."

"Very good, sir."

The Captain flipped the communicator switch back, and pulled over the announcement mike.

"Attention, please. You may release all safety-belts and proceed as normal. Correction of course has been postponed. You will be warned before it is resumed. That is all."

Captain and navigator looked at one another again. Their faces were grave, and their eyes troubled. . . .

Captain Winters studied his audience. It comprised everyone aboard the *Falcon*. Fourteen men and one woman. Six of the men were his crew; the rest passengers. He watched them as they found themselves places in the ship's small living-room. He would have been happier if his cargo had consisted of more freight and fewer passengers. Passengers, having nothing to occupy them, were always making mischief one way and another. Moreover, it was not a quiet, subservient type of man who recommended himself for a job as a miner, prospector, or general adventurer on Mars.

The woman could have caused a great deal of trouble aboard had she been so minded. Luckily she was diffident, self-effacing. But even though at times she was irritatingly without spirit, he thanked his luck that she had not turned out to be some incendiary blonde who would only add to his troubles.

All the same, he reminded himself, regarding her as she sat beside her husband, she could not be quite as meek as she looked. Carter must have been right when he spoke of a stiffening motive somewhere—without that she could never have started on the journey at all, and she would certainly not be coming through steadfast and uncomplaining so far. He glanced at the woman's husband. Queer creatures, women. Morgan was

Survival

all right, but there was nothing about him, one would have said, to lead a woman on a trip like this. . . .

He waited until they had finished shuffling around and fitting themselves in. Silence fell. He let his gaze dwell on each face in turn. His own expression was serious.

"Mrs. Morgan and gentlemen," he began. "I have called you here together because it seemed best to me that each of you should have a clear understanding of our present position.

"It is this. Our lateral tubes have failed. They are, for reasons which we have not yet been able to ascertain, useless. In the case of the port laterals they are burnt out, and irreplaceable.

"In case some of you do not know what that implies, I should tell you that it is upon the laterals that the navigation of the ship depends. The main drive tubes give us the initial impetus for take-off. After that they are shut off, leaving us in free fall. Any deviations from the course plotted are corrected by suitable bursts from the laterals.

"But it is not only for steering that we use them. In landing, which is an infinitely more complex job than take-off, they are essential. We brake by reversing the ship and using the main drive to check our speed. But I think you can scarcely fail to realize that it is an operation of the greatest delicacy to keep the huge mass of such a ship as this perfectly balanced upon the thrust of her drive as she descends. It is the laterals which make such balance possible. Without them it cannot be done."

A dead silence held the room for some seconds. Then a voice asked, drawling:

"What you're saying, Captain, is, the way things are, we can neither steer nor land—is that it?"

Captain Winters looked at the speaker. He was a big man. Without exerting himself, and, apparently, without intention, he seemed to possess a natural domination over the rest.

"That is exactly what I mean," he replied.

A tenseness came over the room. There was the sound of a quickly drawn breath here and there.

The man with the slow voice nodded, fatalistically. Someone else asked:

"Does that mean that we might crash on Mars?"

"No," said the Captain. "If we go on traveling as we are now, slightly off course, we shall miss Mars altogether."

"And so go on out to play tag with the asteroids," another voice suggested.

"That is what would happen if we did nothing about it. But there is a way we can stop that, if we can manage it." The Captain paused, aware that he had their absorbed attention. He continued:

"You must all be well aware from the peculiar behavior of space as seen from our ports that we are now tumbling along all as—er—head over heels. This is due to the explosion of the port laterals. It is a highly unorthodox method of traveling, but it does mean that by an impulse from our main tubes given at exactly the critical moment we should be able to alter our course approximately as we require."

"And how much good is that going to do us if we can't land?" somebody wanted to know. The Captain ignored the interruption. He continued:

"I have been in touch by radio with both home and Mars, and have reported our state. I have also informed them that I intend to attempt the one possible course open to me. That is of using the main drive in an attempt to throw the ship into an orbit about Mars.

"If that is successful we shall avoid two dangers—that of shooting on towards the outer parts of the system, and of crashing on Mars. I think we have a good chance of bringing it off."

When he stopped speaking he saw alarm in several faces, thoughtful concentration in others. He noticed Mrs. Morgan holding tightly to her husband's hand, her face a little paler than usual. It was the man with the drawl who broke the silence.

"You *think* there is a good chance?" he repeated questioningly.

"I do. I also think it is the only chance. But I'm not going to try to fool you by pretending complete confidence. It's too serious for that."

"And if we do get into this orbit?"

"They will try to keep a radar fix on us, and send help as soon as possible."

"H'm," said the questioner. "And what do you personally think about that, Captain?"

"I—well, it isn't going to be easy. But we're all in this together, so I'll tell you just what they told me. At the very best we can't expect them to reach us for some months. The ship will have to come from Earth. The two planets are well past conjunction now. I'm afraid it's going to mean quite a wait."

"Can we—hold out long enough, Captain?"

"According to my calculations we should be able to hold out for about seventeen or eighteen weeks."

"And that will be long enough?"

"It'll have to be."

He broke the thoughtful pause that followed by continuing in a brisker manner.

"This is not going to be comfortable, or pleasant. But, if we all play our parts, and keep strictly to the necessary measures, it can be done. Now, there are three essentials: air to breathe—well, luckily we shan't have to worry about that. The regeneration plant and stock of spare cylinders, and cylinders in cargo will look after that for a long time. Water will be rationed. Two pints each every twenty-four hours, for *everything*. Luckily we shall be able to draw water from the fuel tanks, or it would be a great deal less than that. The thing that is going to be our most serious worry is food."

He explained his proposals further, with patient clarity. At the end he added: "And now I expect you have some questions?"

A small, wiry man with a weather-beaten face asked:

"Is there no hope at all of getting the lateral tubes to work again?"

Captain Winters shook his head.

"Negligible. The impellent section of a ship is not constructed to be accessible in space. We shall keep on trying, of course, but even if the others could be made to fire, we should still be unable to repair the port laterals."

He did his best to answer the few more questions that followed in ways that held a balance between easy confidence and despondency. The prospect was by no means good. Before help could possibly reach them they were all going to need all the nerve and resolution they had—and out of sixteen persons some must be weaker than others.

His gaze rested again on Alice Morgan and her husband beside her. Her presence was certainly a possible source of trouble. When it came to the pinch the man would have more strain on account of her—and, most likely, fewer scruples.

Since the woman was here, she must share the consequences equally with the rest. There could be no privilege. In a sharp emergency one could afford a heroic gesture, but preferential treatment of any one person in the long ordeal which they must face would create an impossible situation. Make any allowances for her, and you would be called on to make allowances for others on health or other grounds—with heaven knew what complications to follow.

A fair chance with the rest was the best he could do for her

—not, he felt, looking at her as she clutched her husband's hand and looked at him from wide eyes in a pale face, not a very good best.

He hoped she would not be the first to go under. It would be better for morale if she were not the very first. . . .

She was not the first to go. For nearly three months nobody went.

The *Falcon*, by means of skilfully timed bursts on the main tubes, had succeeded in nudging herself into an orbital relationship with Mars. After that, there was little that the crew could do for her. At the distance of equilibrium she had become a very minor satellite, rolling and tumbling on her circular course, destined, so far as anyone could see, to continue this untidy progress until help reached them or perhaps forever. . . .

Inboard, the complexity of her twisting somersaults was not perceptible unless one deliberately uncovered a port. If one did, the crazy cavortings of the universe outside produced such a sense of bewilderment that one gladly shut the cover again to preserve the illusion of stability within. Even Captain Winters and the Navigating Officer took their observations as swiftly as possible and were relieved when they had shut the whizzing constellations off the screen, and could take refuge in relativity.

For all her occupants the *Falcon* had become a small, independent world, very sharply finite in space, and scarcely less so in time.

It was, moreover, a world with a very low standard of living; a community with short tempers, weakening distempers, aching bellies, and ragged nerves. It was a group in which each man watched on a trigger of suspicion for a hairsbreadth difference in the next man's ration, and where the little he ate so avidly was not enough to quiet the rumblings of his stomach. He was ravenous when he went to sleep; more ravenous when he woke from dreams of food.

Men who had started from Earth full-bodied were now gaunt and lean, their faces had hardened from curved contours into angled planes and changed their healthy colors for a gray pallor in which their eyes glittered unnaturally. They had all grown weaker. The weakest lay on their couches torpidly. The more fortunate looked at them from time to time with a question in their eyes. It was not difficult to read the question: "Why do we go on wasting good food on this guy? Looks like he's

Survival

booked, anyway." But as yet no one had taken up that booking.

The situation was worse than Captain Winters had foreseen. There had been bad stowage. The cans in several cases of meat had collapsed under the terrific pressure of other cans above them during take-off. The resulting mess was now describing an orbit of its own around the ship. He had had to throw it out secretly. If the men had known of it, they would have eaten it gladly, maggots and all. Another case shown on his inventory had disappeared. He still did not know how. The ship had been searched for it without trace. Much of the emergency stores consisted of dehydrated foods for which he dared not spare sufficient water, so that though edible they were painfully unattractive. They had been intended simply as a supplement in case the estimated time was overrun, and were not extensive. Little in the cargo was edible, and that mostly small cans of luxuries. As a result, he had had to reduce the rations expected to stretch meagerly over seventeen weeks. And even so, they would not last that long.

The first who did go owed it neither to sickness nor malnutrition, but to accident.

Jevons, the chief engineer, maintained that the only way to locate and correct the trouble with the laterals was to effect an entry into the propellent section of the ship. Owing to the tanks which backed up against the bulkhead separating the sections this could not be achieved from within the ship herself.

It had proved impossible with the tools available to cut a slice out of the hull; the temperature of space and the conductivity of the hull caused all their heat to run away and dissipate itself without making the least impression on the tough skin. The one way he could see of getting in was to cut away round the burnt-out tubes of the port laterals. It was debatable whether this was worth while since the other laterals would still be unbalanced on the port side, but where he found opposition solidly against him was in the matter of using precious oxygen to operate his cutters. He had to accept that ban, but he refused to relinquish his plan altogether.

"Very well," he said, grimly. "We're like rats in a trap, but Bowman and I aim to do more than just keep the trap going, and we're going to try, even if we have to cut our way into the damned ship by hand."

Captain Winters had okayed that; not that he believed that anything useful would come of it, but it would keep Jevons quiet, and do no one else any harm. So for weeks Jevons and Bowman had got into their spacesuits and worked their shifts. Oblivious after a time of the wheeling heavens about them, they kept doggedly on with their sawing and filing. Their progress, pitifully slow at best, had grown even slower as they became weaker.

Just what Bowman was attempting when he met his end still remained a mystery. He had not confided in Jevons. All that anyone knew about it was the sudden lurch of the ship and the clang of reverberations running up and down the hull. Possibly it was an accident. More likely he had become impatient and laid a small charge to blast an opening.

For the first time for weeks ports were uncovered and faces looked out giddily at the wheeling stars. Bowman came into sight. He was drifting inertly, a dozen yards or more outboard. His suit was deflated, and a large gash showed in the material of the left sleeve.

The consciousness of a corpse floating round and round you like a minor moon is no improver of already lowered morale. Push it away, and it still circles, though at a greater distance. Someday a proper ceremony for the situation would be invented—perhaps a small rocket would launch the poor remains upon their last, infinite voyage. Meanwhile, lacking a precedent, Captain Winters decided to pay the body the decent respect of having it brought inboard. The refrigeration plant had to be kept going to preserve the small remaining stocks of food, but several sections of it were empty. . . .

A day and a night by the clock had passed since the provisional interment of Bowman when a modest knock came on the control room door. The Captain laid blotting-paper carefully over his latest entry in the log, and closed the book.

"Come in," he said.

The door opened just widely enough to admit Alice Morgan. She slipped in, and shut it behind her. He was somewhat surprised to see her. She had kept sedulously in the background, putting the few requests she had made through the intermediation of her husband. He noticed the changes in her. She was haggard now as they all were, and her eyes anxious. She was also nervous. The fingers of her thin hands sought one another and interlocked themselves for confidence. Clearly she was having to push herself to raise whatever was in her mind. He smiled in order to encourage her.

"Come and sit down, Mrs. Morgan," he invited, amiably. She crossed the room with a slight clicking from her magnetic soles, and took the chair he indicated. She seated herself uneasily, and on the forward edge.

It had been sheer cruelty to bring her on this voyage, he reflected again. She had been at least a pretty little thing, now she was no longer that. Why couldn't that fool husband of hers have left her in her proper setting—a nice quiet suburb, a gentle routine, a life where she would be protected from exaction and alarm alike. It surprised him again that she had had the resolution and the stamina to survive conditions on the *Falcon* as long as this. Fate would probably have been kinder to her if it had disallowed that. He spoke to her quietly, for she perched rather than sat, making him think of a bird ready to take off at any sudden movement.

"And what can I do for you, Mrs. Morgan?"

Alice's fingers twined and intertwined. She watched them doing it. She looked up, opened her mouth to speak, closed it again.

"It isn't very easy," she murmured apologetically.

Trying to help her, he said:

"No need to be nervous, Mrs. Morgan. Just tell me what's on your mind. Has one of them been—bothering you?"

She shook her head.

"Oh, no, Captain Winters. It's nothing like that at all."

"What is it, then?"

"It's—it's the rations, Captain. I'm not getting enough food."

The kindly concern froze out of his face.

"None of us is," he told her, shortly.

"I know," she said, hurriedly. "I know, but—"

"But what?" he inquired in a chill tone.

She drew a breath.

"There's the man who died yesterday. Bowman. I thought if I could have his rations—"

The sentence trailed away as she saw the expression on the Captain's face.

He was not acting. He was feeling just as shocked as he looked. Of all the impudent suggestions that ever had come his way, none had astounded him more. He gazed dumbfounded at the source of the outrageous proposition. Her eyes met his, but, oddly, with less timidity than before. There was no sign of shame in them.

"I've *got* to have more food," she said, intensely.

Captain Winters' anger mounted.

"So you thought you'd just snatch a dead man's share as well as your own! I'd better not tell you in words just where I class that suggestion, young woman. But you can understand this: we share, and we share equally. What Bowman's death means to us is that we can keep on having the same ration for a little longer—that, and only that. And now I think you had better go."

But Alice Morgan made no move to go. She sat there with her lips pressed together, her eyes a little narrowed, quite still save that her hands trembled. Even through his indignation the Captain felt surprise, as though he had watched a hearth cat suddenly become a hunter. She said stubbornly:

"I haven't asked for any privilege until now, Captain. I wouldn't ask you now if it weren't absolutely necessary. But that man's death gives us a margin now. And I *must* have more food."

The Captain controlled himself with an effort.

"Bowman's death has *not* given us a margin, or a windfall—all it has done is to extend by a day or two the chance of our survival. Do you think that every one of us doesn't ache just as much as you do for more food? In all my considerable experience of effrontery—"

She raised her thin hand to stop him. The harness of her eyes made him wonder why he had ever thought her timid.

"Captain. Look at me!" she said, in a harsh tone.

He looked. Presently his expression of anger faded into shocked astonishment. A faint tinge of pink stole into her pale cheeks.

"Yes," she said. "You see, you've *got* to give me more food. My baby *must* have the chance to live."

The Captain continued to stare at her as if mesmerized. Presently he shut his eyes, and passed his hand over his brow.

"God in heaven. This is terrible," he murmured.

Alice Morgan said seriously, as if she had already considered that very point:

"No. It isn't terrible—not if my baby lives." He looked at her helplessly, without speaking. She went on:

"It wouldn't be robbing anyone, you see. Bowman doesn't need his rations any more—but my baby does. It's quite simple, really." She looked questioningly at the Captain. He had no comment ready. She continued: "So you couldn't call it unfair. After all, I'm two people now, really, aren't I? I *need* more food. If you don't let me have it you will be murdering my baby. So you *must* . . . *must* . . . My baby has *got* to live—he's got to. . . ."

When she had gone Captain Winters mopped his forehead,

unlocked his private drawer, and took out one of his carefully hoarded bottles of whiskey. He had the self-restraint to take only a small pull on the drinking-tube and then put it back. It revived him a little, but his eyes were still shocked and worried.

Would it not have been kinder in the end to tell the woman that her baby had no chance at all of being born? That would have been honest; but he doubted whether the coiner of the phrase about honesty being the best policy had known a great deal about group-morale. Had he told her that, it would have been impossible to avoid telling her why, and once she knew why it would have been impossible for her not to confide it, if only to her husband. And then it would be too late.

The Captain opened the top drawer, and regarded the pistol within. There was always that. He was tempted to take hold of it now and use it. There wasn't much use in playing the silly game out. Sooner or later it would have to come to that, anyway.

He frowned at it, hesitating. Then he put out his right hand and gave the thing a flip with his finger, sending it floating to the back of the drawer, out of sight. He closed the drawer. Not yet. . . .

But perhaps he had better begin to carry it soon. So far, his authority had held. There had been nothing worse than safety-valve grumbling. But a time would come when he was going to need the pistol either for them or for himself.

If they should begin to suspect that the encouraging bulletins that he pinned up on the board from time to time were fakes; if they should somehow find out that the rescue ship which they believed to be hurtling through space towards them had not, in fact, even yet been able to take off from Earth—that was when hell would start breaking loose.

It might be safer if there were to be an accident with the radio equipment before long. . . .

"Taken your time, haven't you?" Captain Winters asked. He spoke shortly because he was irritable, not because it mattered in the least how long anyone took over anything now.

The Navigating Officer made no reply. His boots clicked across the floor. A key and an identity bracelet drifted towards the Captain, an inch or so above the surface of his desk. He put out a hand to check them.

"I—" he began. Then he caught sight of the other's face. "Good God, man, what's the matter with you?"

He felt some compunction. He wanted Bowman's identity bracelet for the record, but there had been no real need to send

Carter for it. A man who had died Bowman's death would be a piteous sight. That was why they had left him still in his spacesuit instead of undressing him. All the same, he had thought that Carter was tougher stuff. He brought out a bottle. The last bottle.

"Better have a shot of this," he said.

The navigator did, and put his head in his hands. The Captain carefully rescued the bottle from its mid-air drift, and put it away. Presently the Navigating Officer said, without looking up:

"I'm sorry, sir."

"That's okay, Carter. Nasty job. Should have done it myself."

The other shuddered slightly. A minute passed in silence while he got a grip on himself. Then he looked up and met the Captain's eyes.

"It—it wasn't just that, sir."

The Captain looked puzzled.

"How do you mean?" he asked.

The officer's lips trembled. He did not form his words properly, and he stammered.

"Pull yourself together. What are you trying to say?" The Captain spoke sharply to stiffen him.

Carter jerked his head slightly. His lips stopped trembling.

"He—he—" he floundered; then he tried again, in a rush. "He—hasn't any legs, sir."

"Who? What *is* this? You mean Bowman hasn't any legs?"

"Y-yes, sir."

"Nonsense, man. I was there when he was brought in. So were you. He had legs, all right."

"Yes, sir. He did have legs then—but he hasn't now!"

The Captain sat very still. For some seconds there was no sound in the control-room but the clicking of the chronometer. Then he spoke with difficulty, getting no further than two words:

"You mean—?"

"What else could it be, sir?"

"God in heaven!" gasped the Captain.

He sat staring with eyes that had taken on the horror that lay in the other man's. . . .

Two men moved silently, with socks over their magnetic soles. They stopped opposite the door of one of the refrigeration compartments. One of them produced a slender key. He slipped it into the lock, felt delicately with it among the wards for a moment, and then turned it with a click. As the door swung open a pistol fired twice from within the refrigerator. The man who was pulling the door sagged at the knees, and hung in mid-air.

The other man still was behind the half-opened door. He snatched a pistol from his pocket and slid it swiftly round the corner of the door, pointing into the refrigerator. He pulled the trigger twice.

A figure in a spacesuit launched itself out of the refrigerator, sailing uncannily across the room. The other man shot at it as it swept past him. The spacesuited figure collided with the opposite wall, recoiled slightly, and hung there. Before it could turn and use the pistol in its hand, the other man fired again. The figure jerked, and floated back against the wall. The man kept his pistol trained, but the spacesuit swayed there, flaccid and inert.

The door by which the men had entered opened with a sudden clang. The Navigating Officer on the threshold did not hesitate. He fired slightly after the other, but he kept on firing. . . .

When his pistol was empty the man in front of him swayed queerly, anchored by his boots; there was no other movement in him. The Navigating Officer put out a hand and steadied himself by the doorframe. Then, slowly and painfully, he made his way across to the figure in the spacesuit. There were gashes in the suit. He managed to unlock the helmet and pull it away.

The Captain's face looked somewhat grayer than undernourishment had made it. His eyes opened slowly. He said in a whisper:

"Your job now, Carter. Good luck!"

The Navigating Officer tried to answer, but there were no words, only a bubbling of blood in his throat. His hands relaxed. There was a dark stain still spreading on his uniform. Presently his body hung listlessly swaying beside his Captain's.

"I figured they were going to last a lot longer than this," said the small man with the sandy mustache.

The man with the drawl looked at him steadily.

"Oh, you did, did you? And do you reckon your figuring's reliable?"

The smaller man shifted awkwardly. He ran the tip of his tongue along his lips.

"Well, there was Bowman. Then those four. Then the two that died. That's seven."

"Sure. That's seven. Well?" inquired the big man softly. He

was not as big as he had been, but he still had a large frame. Under his intent regard the emaciated small man seemed to shrivel a little more.

"Er—nothing. Maybe my figuring was kind of hopeful," he said.

"Maybe. My advice to you is to quit figuring and keep on hoping. Huh?"

The small man wilted. "Er—yes. I guess so."

The big man looked round the living-room, counting heads. "Okay. Let's start," he said.

A silence fell on the rest. They gazed at him with uneasy fascination. They fidgeted. One or two nibbled at their finger nails. The big man leaned forward. He put a space-helmet, inverted, on the table. In his customary leisurely fashion he said:

"We shall draw for it. Each of us will take a paper and hold it up unopened until I give the word. *Un*opened. Got that?"

They nodded. Every eye was fixed intently upon his face.

"Good. Now one of those pieces of paper in the helmet is marked with a cross. Ray, I want you to count the pieces there and make sure that there are nine—"

"Eight!" said Alice Morgan's voice, sharply.

All the heads turned towards her as if pulled by strings. The faces looked startled, as though the owners might have heard a turtle-dove roar. Alice sat embarrassed under the combined gaze, but she held herself steady and her mouth was set in a straight line. The man in charge of the proceedings studied her.

"Well, well," he drawled. "So you don't want to take a hand in our little game!"

"No," said Alice.

"You've shared equally with us so far—but now we have reached this regrettable stage you don't want to?"

"No," agreed Alice again.

He raised his eyebrows.

"You are appealing to our chivalry, perhaps?"

"No," said Alice once more. "I'm denying the equity of what you call your game. The one who draws the cross dies—isn't that the plan?"

"Pro bono publico," said the big man. "Deplorable, of course, but unfortunately necessary."

"But if *I* draw it, two must die. Do you call that equitable?" Alice asked.

The group looked taken aback. Alice waited.

The big man fumbled it. For once he was at a loss.

"Well," said Alice, "isn't that so?"

One of the others broke the silence to observe: "The question of the exact stage when the personality, the soul of the individual, takes form is still highly debatable. Some have held that until there is separate existence—"

The drawling voice of the big man cut him short. "I think we can leave that point to the theologians, Sam. This is more in the Wisdom of Solomon class. The point would seem to be that Mrs. Morgan claims exemption on account of her condition."

"My baby has a right to live," Alice said doggedly.

"We all have a right to live. We all want to live," someone put in.

"Why should you—?" another began; but the drawling voice dominated again:

"Very well, gentlemen. Let us be formal. Let us be democratic. We will vote on it. The question is put: do you consider Mrs. Morgan's claim to be valid—or should she take her chance with the rest of us? Those in—"

"Just a minute," said Alice, in a firmer voice than any of them had heard her use. "Before you start voting on that you'd better listen to me a bit." She looked round, making sure she had the atttention of all of them. She had; and their astonishment as well.

"Now the first thing is that I am a lot more important than any of you," she told them simply. "No, you needn't smile. I am—and I'll tell you why.

"Before the radio broke down—"

"Before the Captain wrecked it, you mean," someone corrected her.

"Well, before it became useless," she compromised, "Captain Winters was in regular touch with home. He gave them news of us. The news that the Press wanted most was about me. Women, particularly women in unusual situations, are always news. He told me I was in the headlines: GIRL-WIFE IN DOOM ROCKET, WOMAN'S SPACEWRECK ORDEAL, that sort of thing. And if you haven't forgotten how newspapers look, you can imagine the leads, too: 'Trapped in their living space tomb, a girl and fifteen men now wheel helplessly around the planet Mars . . .'

"All of you are just men—hulks, like the ship. I am a woman, therefore my position is romantic, so I am young, glamorous, beautiful . . ." Her thin face showed for a moment the trace of a wry smile. "I am a heroine . . ."

She paused, letting the idea sink in. Then she went on:

"I was a heroine even before Captain Winters told them that I was pregnant. But after that I became a phenomenon. There were demands for interviews. I wrote one, and Captain Winters transmitted it for me. There have been interviews with my parents and my friends, anyone who knew me. And now an enormous number of people know a great deal about me. They are intensely interested in me. They are even more interested in my baby—which is likely to be the first baby ever born in a spaceship . . .

"Now do you begin to see? You have a fine tale ready. Bowman, my husband, Captain Winters and the rest were heroically struggling to repair the port laterals. There was an explosion. It blew them all away out into space.

"You may get away with that. But if there is no trace of me and my baby—or of our bodies—*then* what are you going to say? How will you explain that?"

She looked round the faces again.

"Well, what *are* you going to say? That I, too, was outside repairing the port laterals? That I committed suicide by shooting myself out into space with a rocket?

"Just think it over. The whole world's press is wanting to know about me—with all the details. It'll have to be a mighty good story to stand up to that. And if it doesn't stand up—well, the rescue won't have done much good.

"You'll not have a chance in hell. You'll hang, or you'll fry, every one of you—unless it happens they lynch you first. . . ."

There was silence in the room as she finished speaking. Most of the faces showed the astonishment of men ferociously attacked by a Pekinese, and at a loss for suitable comment.

The big man sat sunk in reflection for a minute or more. Then he looked up, rubbing the stubble on his sharp-boned chin thoughtfully. He glanced round the others and then let his eyes rest on Alice. For a moment there was a twitch at the corner of his mouth.

"Madam," he drawled, "you are probably a great loss to the legal profession." He turned away. "We shall have to reconsider this matter before our next meeting. But, for the present, Ray, *eight* pieces of paper as the lady said. . . ."

"It's her!" said the Second, over the Skipper's shoulder.

The Skipper moved irritably. "Of course it's her. What else'd you expect to find whirling through space like a sozzled owl?" He studied the screen for a moment. "Not a sign. Every port covered."

"Do you think there's a chance, Skipper?"

"What, after all this time! No, Tommy, not a ghost of it. We're—just the morticians, I guess."

Survival 141

"How'll we get aboard her, Skip?"

The Skipper watched the gyration of the *Falcon* with a calculating eye.

"Well, there aren't any rules, but I reckon if we can get a cable on her we *might* be able to play her gently, like a big fish. It'll be tricky, though."

Tricky it was. Five times the magnet projected from the rescue ship failed to make contact. The sixth attempt was better judged. When the magnet drifted close to the *Falcon* the current was switched on for a moment. It changed course, and floated nearer to the ship. When it was almost in contact the switch went over again. It darted forward, and glued itself limpet-like to the hull.

Then followed the long game of playing the *Falcon;* of keeping tension on the cable between the two ships, but not too much tension, and of holding the rescue ship from being herself thrown into a roll by the pull. Three times the cable parted, but at last, after weary hours of adroit maneuver by the rescue ship the derelict's motion had been reduced to a slow twist. There was still no trace of life aboard. The rescue ship closed a little.

The Captain, the Third Officer and the doctor fastened on their spacesuits and went outboard. They made their way forward to the winch. The Captain looped a short length of line over the cable, and fastened both ends of it to his belt. He laid hold of the cable with both hands, and with a heave sent himself skimming into space. The others followed him along the guiding cable.

They gathered beside the *Falcon's* entrance port. The Third Officer took a crank from his satchel. He inserted it in an opening, and began to turn until he was satisfied that the inner door of the airlock was closed. When it would turn no more, he withdrew it, and fitted it into the next opening; that should set the motors pumping air out of the lock—if there were air, and if there were still current to work the motors. The Captain held a microphone against the hull, and listened. He caught a humming.

"Okay. They're running," he said.

He waited until the humming stopped.

"Right. Open her up," he directed.

The Third Officer inserted his crank again, and wound it. The main port opened inwards, leaving a dark gap in the shining hull. The three looked at the opening somberly for some seconds. With a grim quietness the Captain's voice said: "Well. Here we go!"

They moved carefully and slowly into the blackness, listening.

The Third Officer's voice murmured:

> *"The silence that is in the starry sky,*
> *The sleep that is among the lonely hills . . ."*

Presently the Captain's voice asked:
"How's the air, Doc?"

The doctor looked at his gauges.

"It's okay," he said, in some surprise. "Pressure's about six ounces down, that's all." He began to unfasten his helmet. The others copied him. The Captain made a face as he took his off.

"The place stinks," he said, uneasily. "Let's—get on with it."

He led the way towards the lounge. They entered it apprehensively.

The scene was uncanny and bewildering. Though the gyrations of the *Falcon* had been reduced, every loose object in her continued to circle until it met a solid obstruction and bounced off it upon a new course. The result was a medley of wayward items churning slowly hither and thither.

"Nobody here, anyway," said the Captain, practically. "Doc, do you think—?"

He broke off at the sight of the doctor's strange expression. He followed the line of the other's gaze. The doctor was looking at the drifting flotsam of the place. Among the flow of books, cans, playing-cards, boots and miscellaneous rubbish, his attention was riveted upon a bone. It was large and clean and had been cracked open.

The Captain nudged him. "What's the matter, Doc?"

The doctor turned unseeing eyes upon him for a moment, and then looked back at the drifting bone.

"That—" he said in an unsteady voice—"that, Skipper, is a human femur."

In the long moment that followed while they stared at the grisly relic the silence which had lain over the *Falcon* was broken. The sound of a voice rose, thin, uncertain, but perfectly clear. The three looked incredulously at one another as they listened:

> *"Rock-a-bye baby*
> *On the tree top*
> *When the wind blows*
> *The cradle will rock . . ."*

Alice sat on the side of her bunk, swaying a little, and holding her baby to her. It smiled, and reached up one miniature hand to pat her cheek as she sang:

> ". . . *When the bough breaks*
> *The cradle will fall.*
> *Down will—*"

Her song cut off suddenly at the click of the opening door. For a moment she stared as blankly at the three figures in the opening as they at her. Her face was a mask with harsh lines drawn from the points where the skin was stretched tightly over prominent bones. Then a trace of expression came over it. Her eyes brightened. Her lips curved in a travesty of a smile.

She loosed her arms from about the baby, and it hung there in mid-air, chuckling a little to itself. She slid her right hand under the pillow of the bunk and drew it out again, holding a pistol.

The black shape of the pistol looked enormous in her transparently thin hand as she pointed it at the men who stood transfixed in the doorway.

"Look, baby," she said. "Look there. Food. Lovely food . . ."

MACHINE

By John W. Jakes

"Helen, I want you to get rid of that Goddamned toaster!" Charlie shouted, nursing his hand and glaring at the shining silver box buzzing faintly beside the remains of breakfast.

His wife, looking fresh and pretty in her print robe, hurried into the kitchen, pouting a little as she said, "Charlie, I wish you wouldn't shout so. The neighbors will hear you. What's the trouble?"

Charlie sat down and fumbled for a cigarette. He pointed at the toaster and glowered, "I reached out to put another piece of bread in, and the thing jumped and burned my hand."

"Oh, Charlie," Helen cooed, like a mother reproving a naughty boy.

"I swear to God that's what happened," Charlie said earnestly, showing her his hand, with a small area of skin colored a bright pink.

Helen patted his arm. "Aunt Bertha gave us that toaster and it's very useful."

"I burned myself last week, too. I didn't tell you about that."

Helen sank down into a chair.

"I don't know what we're going to do, Charlie. Your notions about mechanical things are wearing me out." Her voice grew harsh, fingernail-on-blackboard. "Those fixations of your are . . . well, just plain silly."

Charlie rubbed his pink hand. "OK, I don't feel like arguing. We'll decide tonight."

"But I go to the Women's Club tonight. Some very important man is lecturing on psychiatry . . ."

"Gimme a kiss, I gotta leave."

Petulantly Helen kissed him. She couldn't help hugging him a little, too. She did love him.

He hurried out, smiling just a bit. She looked at his hand as

Copyright, 1952, by Fantasy House, Inc., and originally published in *The Magazine of Fantasy and Science Fiction*, April, 1952.

it pulled the doorknob and shut the door. A flash of pink-singed flesh . . .

"He probably bumped against it," she smiled, "the big oaf . . ."

Buzzzz, said the toaster complacently.

Charlie took the streetcar to work. He always rode the trolley because you couldn't trust an automobile. That was part of it.

As the bell jangled and he settled into his seat he thought about Helen, and couldn't see how she could be so blind about the toaster.

I know machines do have souls! he told himself as he had done so many times before. Helen and all the rest laugh, but none of them has ever seen a soul. How can they say a machine doesn't have one, if they don't know what to look for? And then they ignore the evidence that proves—*proves!*—that all mechanical things have souls, some good, some bad, just like men are good and bad. People don't pay any attention to the automobiles that run well for years, or the ones that break down and kill their drivers on the first thousand miles. Those wiring circuits that start fires. Or boilers that explode. Creation —man or machine—is soul! Then there was Rudy Bates, my roommate. Never got beyond his freshman year in college. Always laughing. His bright new automobile—smashed up on a bridge two days after he bought it. Rudy with a broken neck and no more laughter. If you look, you can tell the bad machines. Most people just don't look. The good ones won't hurt you. But the bad ones will . . . kill you. *I* watch, and *I* can see the creations of men go to pieces and kill. The machines with the bad souls . . .

Yeah, Charlie thought as the trolley rumbled, yeah, and that toaster is one of the bad ones. I've got to get rid of it before it does any more damage.

The conductor called his stop and Charlie got off. He could always depend on the trolleys. They were good machines. But the toaster . . .

Walking toward the office building, his mind focused:

Tonight, Helen had said. Tonight was her Women's Club meeting. He'd be home alone . . . to take care of the toaster . . . smash it . . .

He had to smash it before it . . . He couldn't think about it.

Helen left at seven that evening, worried. Charlie had a

strange expression on his face. She decided it might be a good thing to come home early from her meeting. Charlie looked tired and several times she caught him staring up at the shelf where the toaster gleamed. It was silly, of course, but she ought to keep an eye on him.

Charlie finished the dishes and pulled down the kitchen blinds.

Walking into the pantry, he took down the chrome-plated machine and set it on the kitchen table. It squatted there, calm, assured.

Charlie got a hammer from a drawer and walked back to the table.

"Now we'll see who's boss," he growled. He slammed the hammer down on the toaster. But the toaster wasn't under it.

It gave a little jump and slid off onto the floor with a bang. It knew he was trying to kill it. Charlie started to sweat.

He mumbled something half groan, half snarl. Then he bent over and picked up the silver box. That was a mistake.

The lights in the room exploded, dimmed, whirled and exploded once more. Charlie felt sick and infinitely weak.

His hands were frozen to the cool metal of the machine . . . he couldn't let go . . . couldn't . . . couldn't . . .

Then he felt something intangible and yet horribly real . . . creeping . . . from the toaster . . . into his hands . . . creeping . . . taking over . . . making over . . . He didn't even have time to scream.

Helen opened the front door at nine-eighteen. It had been awfully hard to leave such a fascinating lecture. But the speaker's topic, all about people with odd delusions, kept reminding her of her husband.

"Darling?" she called from the living-room.

"Here," a voice said from the kitchen.

Helen didn't like the sound of that voice. She hurried.

Charlie was just going out the kitchen door. Helen saw him toss a mass of something flaccid and sticky into the garbage can. Although curiously lacking in frame and substance, it could have been the toaster.

Charlie carefully replaced the top of the garbage can and came back into the kitchen, watching Helen all the while. "I threw the toaster away," he said.

Helen started for the door. "We're going to bring that right back . . ."

She stopped, wincing. Charlie's hands were fastened on her

arms like bands of metal. Charlie had never been very strong . . . and his eyes . . . they looked so strangely . . . bright . . .

"No," he said, and his voice was flat and brassy. "No."

For the first time in her life, Helen actually felt afraid of him. "All right," she murmured.

"Gimme a kiss," he said, smiling then, as if the smile were a repetition of many others, with no emotion in it. But Helen brushed her lips across his, the fear dying away. . . .

Then she stepped back a bit, cocked her head to one side and grinned.

"You look so funny, Charlie. That man who lectured tonight had a word . . . You look . . . unadjusted!"

Charlie shuffled his legs stiffly. "Not any more." He put his arm around her, and his voice held only the slightest shade of remote disinterest. "Not since we got rid of that toaster."

I AM NOTHING

By Eric Frank Russell

DAVID KORMAN rasped, "Send them the ultimatum."

"Yes, sir, but—"

"But what?"

"It may mean war."

"What of it?"

"Nothing, sir." The other sought a way out. "I merely thought—"

"You are not paid to think," said Korman, acidly. "You are paid only to obey orders."

"Of course, sir. Most certainly." Gathering his papers he backed away hurriedly. "I shall have the ultimatum forwarded to Lani at once."

"You better had!" Korman stared across his ornate desk, watched the door close. Then he voiced an emphatic "Bah!"

A lickspittle. He was surrounded by lickspittles, cravens, weaklings. On all sides were the spineless ready to jump to his command, eager to fawn upon him. They smiled at him with false smiles, hastened into pseudo-agreement with every word he uttered, gave him exaggerated respect that served to cover their inward fears.

There was a reason for all this. He, David Korman, was strong. He was strong in the myriad ways that meant full and complete strength. With his broad body, big jowls, bushy brows and hard gray eyes he looked precisely what he was: a creature of measureless power, mental and physical.

It was good that he should be like this. It was a law of Nature that the weak must give way to the strong. A thoroughly sensible law. Besides, this world of Morcine needed a strong man. Morcine was one world in a cosmos full of potential competitors, all of them born of some misty, long-forgotten planet near a lost sun called Sol. Morcine's duty to itself was

Copyright, 1952, by Street & Smith Publication, Inc., U.S.A. and Great Britain, and originally published in *Astounding Science Fiction*, July, 1952.

I Am Nothing

to grow strong at the expense of the weak. Follow the natural law.

His heavy thumb found the button on his desk, pressed it, and he said into the little silver microphone, "Send in Fleet Commander Rogers at once."

There was a knock at the door and he snapped, "Come in." Then, when Rogers had reached the desk, Korman said, "We have sent the ultimatum."

"Really, sir? Do you suppose they'll accept it?"

"Doesn't matter whether they do or don't," Korman declared. "In either event we'll get our own way." His gaze upon the other became challenging. "Is the fleet disposed in readiness exactly as ordered?"

"It is, sir."

"You are certain of that? You have checked it in person?"

"Yes, sir."

"Very well. These are my orders: the fleet will observe the arrival on Lani of the courier bearing our demands. It will allow twenty-four hours for receipt of a satisfactory reply."

"And if one does not come?"

"It will attack one minute later in full strength. Its immediate task will be to capture and hold an adequate ground base. Having gained it, reinforcements will be poured in and the territorial conquest of the planet can proceed."

"I understand, sir." Rogers prepared to leave. "Is there anything more?"

"Yes," said Korman. "I have one other order. When you are about to seize this base my son's vessel must be the first to land upon it."

Rogers blinked and protested nervously, "But, sir, as a young lieutenant he commands a small scout bearing twenty men. Surely one of our major battleships should be—"

"My son lands first!" Standing up, Korman leaned forward over his desk. His eyes were cold. "The knowledge that Reed Korman, my only child, was in the forefront of the battle will have an excellent psychological effect upon the ordinary masses here. I give it as my order."

"What if something happens?" murmured Rogers, aghast. "What if he should become a casualty, perhaps be killed?"

"That," Korman pointed out, "will enhance the effect."

"All right, sir." Rogers swallowed and hurried out, his features strained.

Had the responsibility for Reed Korman's safety been placed upon his own shoulders? Or was that character behind the desk genuine in his opportunist and dreadful fatalism? He did not know. He knew only that Korman could not be judged by ordinary standards.

Blank-faced and precise, the police escort stood around while Korman got out of the huge official car. He gave them his usual austere look-over while the chauffeur waited, his hand holding the door open. Then Korman mounted the steps to his home, heard the car door close at the sixth step. Invariably it was the sixth step, never the fifth or seventh.

Inside, the maid waited on the same corner of the carpet, her hands ready for his hat, gloves and cloak. She was stiff and starched and never looked directly at him. Not once in fourteen years had she met him eye to eye.

With a disdainful grunt he brushed past her and went into the dining room, took his seat, studied his wife across a long expanse of white cloth filled with silver and crystal.

She was tall and blond and blue-eyed and once had seemed supremely beautiful. Her willowy slenderness had made him think with pleasure of her moving in his arms with the sinuosity of a snake. Now, her slight curves had gained angularity. Her submissive eyes wore crinkles that were not the marks of laughter.

"I've had enough of Lani," he announced. "We're precipitating a showdown. An ultimatum has been sent."

"Yes, David."

That was what he had expected her to say. He could have said it for her. It was her trade-mark, so to speak; always had been, always would be.

Years ago, a quarter of a century back, he had said with becoming politeness, "Mary, I wish to marry you."

"Yes, David."

She had not wanted it—not in the sense that he had wanted it. Her family had pushed her into the arrangement and she had gone where shoved. Life was like that: the pushers and the pushed. Mary was of the latter class. The fact had taken the spice out of romance. The conquest had been too easy. Korman insisted on conquest but he liked it big. Not small.

Later on, when the proper time had come, he had told her, "Mary, I want a son."

She had arranged it precisely as ordered. No slipups. No presenting him with a fat and impudent daughter by way of hapless obstetrical rebellion. A son, eight pounds, afterward named Reed. He had chosen the name.

A faint scowl lay over his broad face as he said, "Almost certainly it means war."

"Does it, David?"

It came without vibrancy or emotion. Dull-toned, her pale oval features expressionless, her eyes submissive. Now and again he wondered whether she hated him with a fierce, turbulent hatred so explosive that it had to be held in check at all costs. He could never be sure of that. Of one thing he was certain: she feared him and had from the very first.

Everyone feared him. Everyone without exception. Those who did not at first meeting soon learned to do so. He saw to that in one way or another. It was good to be feared. It was an excellent substitute for other emotions one has never had or known.

When a child he had feared his father long and ardently; also his mother. Both of them so greatly that their passing had come as a vast relief. Now it was his turn. That, too, was a natural law, fair and logical. What is gained from one generation should be passed to the next. What is denied should likewise be denied.

Justice.

"Reed's scoutship has joined the fleet in readiness for action."

"I know, David."

His eyebrows lifted. "How do you know?"

"I received a letter from him a couple of hours ago." She passed it across.

He was slow to unfold the stiff sheet of paper. He knew what the first two words would be. Getting it open, he found it upside-down, reversed it and looked.

"Dear Mother."

That was her revenge.

"Mary. I want a son."

So she had given him one—and then taken him away.

Now there were letters, perhaps two in one week or one in two months according to the ship's location. Always they were written as though addressing both, always they contained formal love to both, formal hope that both were keeping well.

But always they began, "Dear Mother."

Never, "Dear Father."

Revenge!

Zero hour came and went. Morcine was in a fever of excitement and preparation. Nobody knew what was happening far out in space, not even Korman. There was a time-lag due to sheer distance. Beamed signals from the fleet took many hours to come in.

The first word went straight to Korman's desk where he posed ready to receive it. It said the Lanians had replied with a protest and what they called an appeal to reason. In accordance with instructions the fleet commander had rejected this as unsatisfactory. The attack was on.

"They plead for reasonableness," he growled. "That means they want us to go soft. Life isn't made for the soft." He threw a glance forward. "Is it?"

"No, sir," agreed the messenger with alacrity.

"Tell Bathurst to put the tape on the air at once."

"Yes, sir."

When the other had gone he switched his midget radio and waited. It came in ten minutes, the long, rolling, grandiloquent speech he'd recorded more than a month before. It played on two themes: righteousness and strength, especially strength.

The alleged causes of the war were elucidated in detail, grimly but without ire. That lack of indignation was a telling touch because it suggested the utter inevitability of the present situation and the fact that the powerful have too much justified self-confidence to emote.

As for the causes, he listened to them with boredom. Only the strong know there is but one cause of war. All the other multitudinous reasons recorded in the history books were not real reasons at all. They were nothing but plausible pretexts. There was but one root-cause that persisted right back to the dim days of the jungle. When two monkeys want the same banana, that is war.

Of course, the broadcasting tape wisely refrained from putting the issue so bluntly and revealingly. Weak stomachs require pap. Red meat is exclusively for the strong. So the great antennae of the world network comported themselves accordingly and catered for the general dietary need.

After the broadcast had finished on a heartening note about Morcine's overwhelming power, he leaned back in his chair and thought things over. There was no question of bombing Lani into submission from the upper reaches of its atmosphere. All its cities cowered beneath bombproof hemispherical force fields. Even if they had been wide open he would not have ordered their destruction. It is empty victory to win a few mounds of rubble.

He'd had enough of empty victories. Instinctively, his gray eyes strayed toward the bookcase on which stood the photograph he seldom noticed and then no more than absently. For years it had been there, a subconsciously-observed, taken-for-granted object like the inkpot or radiant heat panel, but less useful than either.

She wasn't like her picture now. Come to think of it, she hadn't been really like it *then*. She had given him obedience and fear before he had learned the need for these in lieu of other needs. At that time he had wanted something else that had not been forthcoming. So long as he could remember, to his very earliest years, it had never been forthcoming, not from anyone, never, never, never.

He jerked his mind back to the subject of Lani. The location of that place and the nature of its defenses determined the pattern of conquest. A ground base must be won, constantly replenished with troops, arms and all auxiliary services. From there the forces of Morcine must expand and, bit by bit, take over all unshielded territory until at last the protected cities stood alone in fateful isolation. The cities would then be permitted to sit under their shields until starved into surrender.

Acquisition of enemy territory was the essential aim. This meant that despite space-going vessels, force shields and all the other redoubtable gadgets of ultra-modernism, the ordinary foot soldier remained the final arbiter of victory. Machines could assault and destroy. Only men could take and hold.

Therefore this was going to be no mere five-minute war. It would run on for a few months, perhaps even a year, with spasms of old style land-fighting as strong points were attacked and defended. There would be bombing perforce limited to road blocks, strategic junctions, enemy assembly and regrouping areas, unshielded but stubborn villages.

There would be some destruction, some casualties. But it was better that way. Real conquest comes only over real obstacles, not imaginary ones. In her hour of triumph Morcine would be feared. Korman would be feared. The feared are respected and that is proper and decent.

If one can have nothing more.

Pictorial records in full color and sound came at the end of the month. Their first showing was in the privacy of his own home to a small audience composed of himself, his wife, a group of government officials and assorted brass hats.

Unhampered by Lanian air defenses, weak from the beginning and now almost wiped out, the long black ships of Morcine dived into the constantly widening base and unloaded great quantities of supplies. Troops moved forward against tough but spasmodic opposition, a growing weight of armored and motorized equipment going with them.

The recording camera trundled across an enormous bridge with thick girders fantastically distorted and with great gaps temporarily filled in. It took them through seven battered villages which the enemy had either defended or given cause to believe they intended to defend. There were shots of crater-pocked roads, skeletal houses, a blackened barn with a swollen horse lying in a field nearby.

And an action-take of an assault on a farmhouse. A patrol, suddenly fired on, dug in and radioed back. A monster on huge, noisy tracks answered their call, rumbled laboriously to within four hundred yards of the objective, spat violently and lavishly from its front turret. A great splash of liquid fell on the farmhouse roof, burst into roaring flame. Figures ran out, seeking cover of an adjacent thicket. The sound track emitted rattling noises. The figures fell over, rolled, jerked, lay still.

The reel ended and Korman said, "I approve it for public exhibition." Getting out of his seat, he frowned around, added, "I have one criticism. My son has taken command of a company of infantry. He is doing a job, like any other man. Why wasn't he featured?"

"We would not depict him except with your approval, sir," said one.

"I not only approve—I order it. Make sure that he is shown next time. Not predominantly. Just sufficiently to let the people see for themselves that he is there, sharing the hardships and the risks."

"Very well, sir."

They packed up and went away. He strolled restlessly on the thick carpet in front of the electric radiator.

"Do them good to know Reed is among those present," he insisted.

"Yes, David." She had taken up some knitting, her needles going *click-click*.

"He's my son."

"Yes, David."

Stopping his pacing, he chewed his bottom lip with irritation. "Can't you say anything but that?"

She raised her eyes. "Do you wish me to?"

"Do I wish!" he echoed. His fists were tight as he resumed his movements to and fro while she returned to her needles.

What did she know of wishes?

What does anyone know?

By the end of four months the territorial grip on Lani had grown to one thousand square miles while men and guns continued to pour in. Progress had been slower than expected. There had been minor blunders at high level, a few of the unforeseeable difficulties that invariably crop up when fighting at long range, and resistance had been desperate where least expected. Nevertheless, progress was being made. Though a little post-dated, the inevitable remained inevitable.

Korman came home, heard the car door snap shut at the sixth step. All was as before except that now a part of the populace insisted on assembling to cheer him indoors. The maid waited, took his things. He stumped heavily to the inner room.

"Reed is being promoted to captain."

She did not answer.

Standing squarely before her, he demanded. "Well, aren't you interested?"

"Of course, David." Putting aside her book, she folded long, thin-fingered hands, looked toward the window.

"What's the matter with you?"

"The matter?" The blond eyebrows arched as her eyes came up. "Nothing is the matter with me. Why do you ask?"

"I can tell." His tones harshened a little. "And I can guess. You don't like Reed being out there. You disapprove of me sending him away from you. You think of him as your son and not mine. You—"

She faced him calmly. "You're rather tired, David. And worried."

"I am not tired," he denied with unnecessary loudness. "Neither am I worried. It is the weak who worry."

"The weak have reason."

"I haven't."

"Then you're just plain hungry." She took a seat at the table. "Have something to eat. It will make you feel better."

Dissatisfied and disgruntled, he got through his evening meal. Mary was holding something back, he knew that with the sureness of one who had lived with her for half his lifetime. But he did not have to force it out of her by autocratic methods. When and only when he had finished eating she surrendered her secret voluntarily. The way in which she did it concealed the blow to come.

"There has been another letter from Reed."

"Yes?" He fingered a glass of wine, felt soothed by food but reluctant to show it. "I know he's happy, healthy and in one piece. If anything went wrong, I'd be the first to learn of it."

"Don't you want to see what he says?" She took it from a little walnut bureau, offered it to him.

He eyed it without reaching for it. "Oh, I suppose it's all the usual chitchat about the war."

"I think you ought to read it," she persisted.

"Do you?" Taking it from her hand he held it unopened, surveyed her curiously. "Why should this particular missive call for my attention? Is it any different from the others? I know without looking that it is addressed to you. Not to me. To you! Never in his life has Reed written a letter specifically to me."

"He writes to both of us."

"Then why can't he start, 'Dear Father and Mother'?"

"Probably it just hasn't occurred to him that you would feel touchy about it. Besides, it's cumbersome."

"Nonsense!"

"Well, you might as well look at it as argue about it unread. You'll have to know sooner or later."

That last remark stimulated him into action. Unfolding it, he grunted as he noted the opening words, then went through ten paragraphs descriptive of war service on another planet. It was the sort of stuff every fighting man sent home. Nothing special about it. Turning the page, he perused the brief remainder. His face went taut and heightened in color.

"Better tell you I've become the willing slave of a Lanian girl. Found her in what little was left of the village of Bluelake which had taken a pretty bad beating from our heavies. She was all alone and, as far as I could discover, seemed to be the sole survivor. Mom, she's got nobody. I'm sending her home on the hospital ship *Istar*. The captain jibbed but dared not refuse a Korman. Please meet her for me and look after her until I get back."

Flinging it onto the table, he swore lengthily and with vim, finishing, "The young imbecile."

Saying nothing, Mary sat watching him, her hands clasped together.

"The eyes of a whole world are on him," he raged. "As a public figure, as the son of his father, he is expected to be an example. And what does he do?"

She remained silent.

"Becomes the easy victim of some designing little skirt who is quick to play upon his sympathies. An enemy female!"

"She must be pretty," said Mary.

"*No* Lanians are pretty," he contradicted in what came near to a shout. "Have you taken leave of your senses?"

"No, David, of course not."

"Then why make such pointless remarks? One idiot in the family is enough." He punched his right fist several times into the palm of his left hand. "At the very time when anti-Lanian sentiment is at its height I can well imagine the effect on public opinion if it became known that we were harboring a specially favored enemy alien, pampering some painted and powdered huzzy who has dug her claws into Reed. I can see her mincing proudly around, one of the vanquished who became a victor by making use of a dope. Reed must be out of his mind."

"Reed is twenty-three," she observed.

"What of it? Are you asserting that there's a specific age at which a man has a right to make a fool of himself?"

"David, I did not say that."

"You implied it." More hand punching. "Reed has shown an unsuspected strain of weakness. It doesn't come from me."

"No, David, it doesn't."

He stared at her, seeking what lay unspoken behind that remark. It eluded him. His mind was not her mind. He could not think in her terms. Only in his own.

"I'll bring this madness to a drastic stop. If Reed lacks strength of character, it is for me to provide it." He found the telephone, remarked as he picked it up, "There are thousands of girls on Morcine. If Reed feels that he must have romance, he can find it at home."

"He's not home," Mary mentioned. "He is far away."

"For a few months. A mere nothing." The phone whirred and he barked into it. "Has the *Istar* left Lani yet?" He held on a while, then racked the instrument and rumbled aggrievedly, "I'd have had her thrown off but it's too late. The *Istar* departed soon after the mailboat that brought his letter." He made a face and it was not pleasant. "The girl is due here tomorrow. She's got a nerve, a blatant impudence. It reveals her character in advance."

Facing the big, slow-ticking clock that stood by the wall, he gazed at it as if tomorrow were due any moment. His mind was working on the problem so suddenly dumped in his lap. After a while he spoke again.

"That scheming baggage is not going to carve herself a comfortable niche in my home, no matter what Reed thinks of her. I will not have her, see?"

"I see, David."

"If he is weak, I am not. So when she arrives I'm going to give her the roughest hour of her life. By the time I've finished she'll be more than glad of passage back to Lani on the next ship. She'll get out in a hurry and for keeps."

Mary remained quiet.

"But I'm not going to indulge a sordid domestic fracas in

public. I won't allow her even the satisfaction of that. I want you to meet her at the spaceport, phone me immediately she arrives, then bring her to my office. I'll cope with her there."

"Yes, David."

"And don't forget to call me beforehand. It will give me time to clear the place and insure some privacy."

"I will remember," she promised.

It was three-thirty in the following afternoon when the call came through. He shooed out a fleet admiral, two generals and an intelligence service director, hurried through the most urgent of his papers, cleared the desk and mentally prepared himself for the distasteful task to come.

In short time his intercom squeaked and his secretary's voice announced, "Two people to see you, sir—Mrs. Korman and Miss Tatiana Hurst."

"Show them in."

He leaned backward, face suitably severe. Tatiana, he thought. An outlandish name. It was easy to visualize the sort of hoyden who owned it: a flouncy thing, aged beyond her years and with a sharp eye to the main chance. The sort who could make easy meat of someone young, inexperienced and impressionable, like Reed. Doubtless she had supreme confidence that she could butter the old man with equal effectiveness and no trouble whatsoever. Hah, that was her mistake.

The door opened and they came in and stood before him without speaking. For half a minute he studied them while his mind did sideslips, repeatedly strove to co-ordinate itself, and a dozen expressions came and went in his face. Finally, he arose slowly to his feet, spoke to Mary, his tones frankly bewildered.

"Well, where is she?"

"This," informed Mary, with unconcealed and inexplicable satisfaction, "is she."

He flopped back into his chair, looked incredulously at Miss Tatiana Hurst. She had skinny legs exposed to knee height. Her clothing was much the worse for wear. Her face was a pale, hollow-cheeked oval from which a pair of enormous dark eyes gazed in a non-focusing, introspective manner as if she continually kept watch within her rather than upon things outside. One small white hand held Mary's, the other arm was around a large and brand new teddy-bear gained from a source at which he could guess. Her age was about eight. Certainly no more than eight.

It was the eyes that got him most, terribly solemn, terribly

I Am Nothing

grave and unwilling to see. There was a coldness in his stomach as he observed them. She was not blind. She could look at him all right—but she looked without really perceiving. The great dark orbs could turn toward him and register the mere essential of his being while all the time they saw only the secret places within herself. It was eerie in the extreme and more than discomforting.

Watching her fascinatedly, he tried to analyze and define the peculiar quality in those optics. He had expected daring, defiance, impudence, passion, anything of which a predatory female was capable. Here, in these radically altered circumstances, one could expect childish embarrassment, self-consciousness, shyness. But she was not shy, he decided. It was something else. In the end he recognized the elusive factor as absentness. She was here yet somehow not with them. She was somewhere else, deep inside a world of her own.

Mary chipped in with a sudden, "Well, David?"

He started at the sound of her voice. Some confusion still cluttered his mind because this culmination differed so greatly from his preconceptions. Mary had enjoyed half an hour in which to accommodate herself to the shock. He had not. It was still fresh and potent.

"Leave her with me for a few minutes," he suggested. "I'll call you when I've finished."

Mary went, her manner that of a woman enjoying something deep and personal. An unexpected satisfaction long overdue.

Korman said with unaccustomed mildness, "Come here, Tatiana."

She moved toward him slowly, each step deliberate and careful, touched the desk, stopped.

"Round this side, please, near to my chair."

With the same almost robotic gait she did as instructed, her dark eyes looking expressionlessly to the front. Arriving at his chair, she waited in silence.

He drew in a deep breath. It seemed to him that her manner was born of a tiny voice insisting, "I must be obedient. I must do as I am told. I can do only what I am told to do."

So she did it as one compelled to accept those things she had no means of resisting. It was surrender to all demands in order to keep one hidden and precious place intact. There was no other way.

Rather appalled, he said, "You're able to speak, aren't you?"

She nodded, slightly and only once.

"But that isn't speech," he pointed out.

There was no desire to contradict or provide proof of ability. She accepted his statement as obvious and left it at that. Silent and immensely grave, she clung to her bear and waited for Korman's world to cease troubling her own.

"Are you glad you're here, or sorry?"

No reaction. Only inward contemplation. Absentness.

"Well, you are glad then?"

A vague half-nod.

"You are not sorry to be here?"

An even vaguer shake.

"Would you rather stay than go back?"

She looked at him, not so much to see him as to insure that he could see her.

He rang his bell, said to Mary, "Take her home."

"Home, David?"

"That's what I said." He did not like the exaggerated sweetness of her tone. It meant something, but he couldn't discern what.

The door closed behind the pair of them. His fingers tapped restlessly on the desk as he pictured those eyes. Something small and bitterly cold was in his insides.

During the next couple of weeks his mind seemed to be filled with more problems than ever before. Like most men of his caliber he had the ability to ponder several subjects at once, but not the insight to detect when one was gaining predominance over the others.

On the first two or three of these days he ignored the pale intruder in his household. Yet he could not deny her presence. She was always there, quiet, obedient, self-effacing, hollow-cheeked and huge-eyed. Often she sat around for long periods without stirring, like a discarded doll.

When addressed by Mary or one of the maids she remained deaf to inconsequential remarks, responded to direct and imperative questions or orders. She would answer with minimum head movements or hand gestures when these sufficed, spoke mono-syllabically in a thin little voice only when speech was unavoidable. During that time Korman did not speak to her at all—but he was compelled to notice her fatalistic acceptance of the fact that she was no part of his complicated life.

After lunch on the fourth day he caught her alone, bent down to her height and demanded, "Tatiana, what is the matter with you? Are you unhappy here?"

One brief shake of her head.

"Then why don't you laugh and play like other—?" He ceased abruptly as Mary entered the room.

"You two having a private gossip?" she inquired.

"As if we could," he snapped.

That same evening he saw the latest pictorial record from the fighting front. It gave him little satisfaction. Indeed, it almost irked him. The zip was missing. Much of the thrill of conquest had mysteriously evaporated from the pictures.

By the end of the fortnight he'd had more than enough of listening for a voice that seldom spoke and meeting eyes that did not see. It was like living with a ghost—and it could not go on. A man is entitled to a modicum of relaxation in his own home.

Certainly he could kick her back to Lani as he had threatened to do at the first. That, however, would be admission of defeat. Korman just could not accept defeat at anyone's hands, much less those of a brooding child. She was not going to edge him out of his own home nor persuade him to throw her out. She was a challenge he had to overcome in a way thoroughly satisfactory to himself.

Summoning his chief scientific adviser to his office, he declaimed with irritation, "Look, I'm saddled with a maladjusted child. My son took a fancy to her and shipped her from Lani. She's getting in my hair. What can be done about it?"

"Afraid I cannot help much, sir."

"Why not?"

"I'm a physicist."

"Well, can you suggest anyone else?"

The other thought a bit, said, "There's nobody in my department, sir. But science isn't solely concerned with production of gadgets. You need a specialist in things less tangible." A pause, then, "The hospital authorities might put you onto someone suitable."

He tried the nearest hospital, got the answer, "A child psychologist is your man."

"Who's the best on this planet?"

"Dr. Jager."

"Contact him for me. I want him at my house this evening, not later than seven o'clock."

Fat, middle-aged and jovial, Jager fell easily into the role of a casual friend who had just dropped in. He chatted a lot of foolishness, included Tatiana in the conversation by throwing odd remarks at her, even held a pretended conversation with her teddy-bear. Twice in an hour she came into his world just long enough to register a fleeting smile—then swiftly she was back in her own.

At the end of this he hinted that he and Tatiana should be left by themselves. Korman went out, convinced that no progress was being or would be made. In the lounge Mary glanced up from her seat.

"Who's our visitor, David? Or is it no business of mine?"

"Some kind of mental specialist. He's examining Tatiana."

"Really?" Again the sweetness that was bitter.

"Yes," he rasped. "Really."

"I didn't think you were interested in her."

"I am not," he asserted. "But Reed is. Now and again I like to remind myself that Reed is my son."

She let the subject drop. Korman got on with some official papers until Jager had finished. Then he went back to the room, leaving Mary immersed in her book. He looked around.

"Where is she?"

"The maid took her. Said it was her bedtime."

"Oh." He found a seat, waited to hear more.

Resting against the edge of a table, Jager explained, "I've a playful little gag for dealing with children who are reluctant to talk. Nine times out of ten it works."

"What is it?"

"I persuade them to *write*. Strangely enough they'll often do that, especially if I make a game of it. I cajole them into writing a story or essay about anything that created a great impression upon them. The results can be very revealing."

"And did you—?"

"A moment, please, Mr. Korman. Before I go further I'd like to impress upon you that children have an inherent ability many authors must envy. They can express themselves with remarkable vividness in simple language, with great economy of words. They create telling effect with what they leave out as much as by what they put in." He eyed Korman speculatively. "You know the circumstances in which your son found this child?"

"Yes, he told us in a letter."

"Well, bearing those circumstances in mind I think you'll find this something exceptional in the way of horror stories." He held out a sheet of paper. "She wrote it unaided." He reached for his hat and coat.

"You're going?" questioned Korman in surprise. "What about your diagnosis? What treatment do you suggest?"

Dr. Jager paused, hand on door. "Mr. Korman, you are an

I Am Nothing

intelligent person." He indicated the sheet the other was holding. "I think that is all you require."

Then he departed. Korman eyed the sheet. It was not filled with words as he'd expected. For a story it was mighty short. He read it.

I am nothing and nobody. My house went bang. My cat was stuck to a wall. I wanted to pull it off. They wouldn't let me. They threw it away.

The cold thing in the pit of his stomach swelled up. He read it again. And again. He went to the base of the stairs and looked up toward where she was sleeping.

The enemy whom he had made nothing.

Slumber came hard that night. Usually he could compose his mind and snatch a nap any time, anywhere, at a moment's notice. Now he was strangely restless, unsettled. His brain was stimulated by he knew not what and it insisted on following tortuous paths.

The frequent waking periods were full of fantastic imaginings wherein he fumbled through a vast and cloying grayness in which was no sound, no voice, no other being. The dreams were worse, full of writhing landscapes spewing smoky columns, with things howling through the sky, with huge, toad-like monsters crawling on metal tracks, with long lines of dusty men singing an aeons-old and forgotten song.

"You've left behind a broken doll."

He awakened early with weary eyes and a tired mind. All morning at the office a multitude of trifling things conspired against him. His ability to concentrate was not up to the mark and several times he had to catch himself on minor errors just made or about to be made. Once or twice he found himself gazing meditatively forward with eyes that did not see to the front but were looking where they had never looked before.

At three in the afternoon his secretary called on the intercom, "Astroleader Warren would like to see you, sir."

"Astroleader?" he echoed, wondering whether he had heard aright. "There's no such title."

"It is a Drakan space-rank."

"Oh, yes, of course. I can tend to him now."

He waited with dull anticipation. The Drakans formed a powerful combine of ten planets at great distance from Morcine. They were so far away that contact came seldom. A battleship of theirs had paid a courtesy call about twice in his lifetime. So this occasion was a rare one.

The visitor entered, a big-built youngster in light-green uniform. Shaking hands with genuine pleasure and great cordiality, he accepted the indicated chair.

"A surprise, eh, Mr. Korman?"

"Very."

"We came in a deuce of a hurry but the trip can't be done in a day. Distance take time unfortunately."

"I know."

"The position is this," explained Warren. "Long while back we received a call from Lani relayed by intervening minor planets. They said they were involved in a serious dispute and feared war. They appealed to us to negotiate as disinterested neutrals."

"Ah, so that's why you've come?"

"Yes, Mr. Korman. We knew the chance was small of arriving in time. There was nothing for it but to come as fast as we could and hope for the best. The role of peacemaker appeals to those with any claim to be civilized."

"Does it?" questioned Korman, watching him.

"It does to us." Leaning forward, Warren met him eye to eye. "We've called at Lani on the way here. They still want peace. They're losing the battle. Therefore we want to know only one thing: Are we too late?"

That was the leading question: Are we too late? Yes or no? Korman stewed it without realizing that not so long ago his answer would have been prompt and automatic. Today, he thought it over.

Yes or no? Yes meant military victory, power and fear. No meant—what? Well, no meant a display of reasonableness in lieu of stubbornness. No meant a considerable change of mind. It struck him suddenly that one must possess redoubtable force of character to throw away a long-nursed viewpoint and adopt a new one. It required moral courage. The weak and the faltering could never achieve it.

"No," he replied slowly. "It is not too late."

Warren stood up, his face showing that this was not the answer he had expected. "You mean, Mr. Korman—?"

"Your journey has not been in vain. You may negotiate."

"On what terms?"

"The fairest to both sides that you can contrive." He switched on his microphone, spoke into it. "Tell Rogers that I order our forces to cease hostilities forthwith. Troops will guard the perimeter of the Lani ground base pending peace negotiations. Citizens of the Drakan Confederation will be permitted unobstructed passage through our lines in either direction."

"Very well, Mr. Korman."

I Am Nothing

Putting the microphone aside, he continued with Warren, "Though far off in mere miles, Lani is near to us as cosmic distances go. It would please me if the Lanians agreed to a union between our planets, with common citizenship, common development of natural resources. But I don't insist upon it. I merely express a wish—knowing that some wishes never come true."

"The notion will be given serious consideration all the same," assured Warren. He shook hands with boyish enthusiasm. "You're a big man, Mr. Korman."

"Am I?" He gave a wry smile. "I'm trying to do a bit of growing in another direction. The original one kind of got used up."

When the other had gone, he tossed a wad of documents into a drawer. Most of them were useless now. Strange how he seemed to be breathing better than ever before, his lungs drawing more fully.

In the outer office he informed them, "It's early yet, but I'm going home. Phone me there if anything urgent comes along."

The chauffeur closed the car door at the sixth step. A weakling, thought Korman as he went into his home. A lamebrain lacking the strength to haul himself out of a self-created rut. One can stay in a rut too long.

He asked the maid, "Where is my wife?"

"Slipped out ten minutes ago, sir. She said she'd be back in half an hour."

"Did she take—"

"No, sir." The maid glanced toward the lounge.

Cautiously he entered the lounge, found the child resting on the settee, head back, eyes closed. A radio played softly nearby. He doubted whether she had turned it on of her own accord or was listening to it. More likely someone else had left it running.

Tiptoeing across the carpet, he cut off the faint music. She opened her eyes, sat upright. Going to the settee, he took the bear from her side and placed it on an arm, positioned himself next to her.

"Tatania," he asked with rough gentleness, "why are you nothing?"

No answer. No change.

"Is it because you have nobody?"

Silence.

"Nobody of your own?" he persisted, feeling a queer kind of desperation. "Not even a kitten?"

She looked down at her shoes, her big eyes partly shielded under pale lids. There was no other reaction.

Defeat. Ah, the bitterness of defeat. It set his fingers fumbling with each other, like those of one in great and unbearable trouble. Phrases tumbled through his mind.

"I am nothing."

"My cat . . . they threw it away."

His gaze wandered blindly over the room while his mind ran round and round her wall of silence seeking a door it could not find. Was there no way in, no way at all?

There was.

He discovered it quite unwittingly.

To himself rather than to her he murmured in a hearable undertone, "Since I was very small I have been surrounded by people. All my life there have been lots of people. But none were mine. Not one was really mine. Not one. I, too, am nothing."

She patted his hand.

The shock was immense. Startled beyond measure, he glanced down at the first touch, watched her give three or four comforting little dabs and hastily withdraw. There was heavy pulsing in his veins. Something within him rapidly became too big to contain.

Twisting sidewise, he snatched her onto his lap, put his arms around her, buried his nose in the soft part of her neck, nuzzled behind her ear, ran his big hand through her hair. And all the time he rocked to and fro with low crooning noises.

She was weeping. She hadn't been able to weep before. She was weeping, not as a woman does, softly and subdued, but like a child, with great racking sobs that she fought hard to suppress.

Her arm was around his neck, tightening, clinging and tightening more while he rocked and stroked and called her "Honey" and uttered silly sounds and wildly extravagant reassurances.

This was victory.

Not empty.

Full.

Victory over self is completely full.

THE BANTAM BOOK SHOPPER
Good Reading Buys for the Entire Family

A magnificent word and picture story of the first fifty years of the American automobile!

Here's a wealth of wonderful memories—new fun and entertainment—for young and old alike, in this glorious album of early automobile pictures and mementos. It recaptures and brings to life the pioneering days of the horseless carriage, with all their humor and excitement.

Treasury of
EARLY AMERICAN AUTOMOBILES
By Floyd Clymer
213 pages, 8 x 11 album size, $5.50
Foreword by James Melton

Over 500 illustrations

This book brings back all the gay, exciting events of that famed era... with its "classic" auto advertisements and cartoons... riotous auto songs that are still sung... famous personalities... dramatic cross country races... and the outstanding cars of yester-year—all brought to you in this big, handsome word and picture album.

You will read realistic, thrilling accounts of the Indianapolis Speedway races, the famous Barney Oldfield in action, the pioneering Glidden Tours, and the unprecedented empire of Henry Ford and other early auto manufacturers.

There are humorous pictures of the famous early cars such as the Stutz Bearcat, the Octoauto, the Duck, with its steering wheel in the back seat, and even the original horseless carriage which mounted a life size horse's head on the radiator to pacify real horses.

SEE THIS BOOK 10 DAYS FREE — JUST MAIL COUPON

McGraw-Hill Book Co.
c/o Bantam Books
Reader Service
25 W. 45 St., N. Y. C.

Send me Clymer's *Treasury of Early American Automobiles* for 10 days' examination on approval. In 10 days I will send $5.50, plus few cents for delivery costs, or return book postpaid. (We pay delivery costs if you remit with coupon; same return privilege.)
(PRINT)

Name..................................

Address...............................

City.............Zone....State....
Canadian price slightly higher

A NEW, AMAZING OFFER—

5 BOOKS FOR $1.00
Save as much as 30c a book— and Bantam pays the postage!

HERE'S YOUR CHANCE to select five bestsellers for only $1.00—books you have always wanted to read. Yes, *any* five books from this list of titles ordinarily selling for 25¢, 35¢ or even 50¢ each! These bestsellers are now in limited supply—books which cost from $2.50 up in their original editions!

This one big chance to save as much as 30¢ a book is offered only to Bantam readers. And your selections will be delivered right to your door—*postage free!* But remember, we have only

Here's how you order. On the coupon at the end of this ad, circle the numbers of the five books you want. Cut out the coupon and enclose it with your remittance in an envelope addressed to Bantam Books, Dept. M1, 657 West Chicago Avenue, Chicago, Illinois.

Minimum order $1. Sorry, no C.O.D.'s because of this special low price. This offer expires Dec. 31, 1955.

a limited supply of these bestselling titles in our Chicago warehouse. So make your choice *now!*

1. THE HOUND OF THE BASKERVILLES, Arthur Conan Doyle. A devilish and savage plot involving a murderous mastiff sends Sherlock Holmes and Dr. Watson on a midnight chase across a terror-ridden moor. 192 pp.

2A. CASTAWAY, James Gould Cozzens. A brilliant novel of overwhelming suspense and horror about a man hiding in a tremendous, deserted department store. By a Pulitzer Prize-winning author. 128 pp.

3. THE WAYWARD BUS, John Steinbeck. The story of a group of men and women stranded on a bus in the California hills—in which John Steinbeck tears the wrapping off so-called "civilized" people. 256 pp.

4. LOUISVILLE SATURDAY, Margaret Long. Powerful, provocative novel of women, alone in an army town, battling the fierce temptation to live only for the moment. 256 pp.

5A. WRITTEN ON THE WIND, Robert Wilder. A scorching novel of

CHOOSE FROM THESE EXCITING BOOKS!

Southern youth—the restless daughter of a wealthy tobacco family and a share-cropper's son.

6. NO SURVIVORS, Will Henry. A renegade cavalry officer, a passionate Indian girl—an account of spectacular courage and blind ferocity, culminating in 1876 at the Little Big Horn.

7A. A ROOM ON THE ROUTE, Godfrey Blunden. A gripping, action-filled story of life in the Soviet—a true-to-life exposure of terror and conspiracy. "A tense, skillfully-told story," declares the Saturday Review. 320 pp.

8. JOHNNY CHRISTMAS, Forrester Blake. Johnny Christmas, who fought his way across the wide, untamed frontier during the Texas-Mexican War, is the epitome of all restless Americans who opened up the West. 288 pp.

9. THERESA, Emile Zola. A powerful story by the famous French novelist of a woman, reckless with desire, and the man she pulled down the path of human degradation. 208 pp.

10. THE BLACK ROSE, Thomas B. Costain. He fought his way to the heart of Kubla Khan's fabulous Oriental empire, then risked his life to save the bewitching slave girl on her way to the Emperor's harem. A best-selling historical novel. 512 pp.

11. WILD IS THE RIVER, Louis Bromfield. When the Yankees seized New Orleans in 1862, Major Tom Bedloe knew how to enjoy the fruits of victory—until he met the Creole Baroness who was as cruel and corrupt as he. 384 pp.

12. STORIES FOR HERE AND NOW, J. Greene & E. Abell. 30 stories of suspense, adventure, love and humor by such famous authors as Stephen Crane, William Faulkner, William Saroyan, James Thurber. 480 pp.

13. B. F.'s DAUGHTER, John P. Marquand. The story of a forceful heiress who tried to live by the ruthless codes her father taught her. A powerful novel by an outstanding novelist. 480 pp.

14. FLAMINGO ROAD, Robert Wilder. A frank novel of vice and political corruption in Florida—and the wily men and willing women who made it possible. 352 pp.

15. A RAGE TO LIVE, John O'Hara. The story of a typical American town —the socialites, and climbers, the crooked politicians and crusaders—and a beautiful and complex woman incapable of remaining faithful to her husband. 672 pp.

16. TIMELESS STORIES FOR TODAY AND TOMORROW, Ed. Ray Bradbury. Twenty-six startling, shocking, frightening science-fantasy adventures by such brilliant writers as Kafka, Steinbeck and Bemelmans. 320 pp.

17. TERROR IN THE STREETS, Harold Whitman. The true and frightening facts about the shocking wave of thrill crimes that is turning our American cities into jungles of terror. 352 pp.

18. BUGLES IN THE AFTERNOON, Ernest Haycox. An unforgettable drama of violence and high courage during the bloody battles with the Sioux over the Western frontier.

19. EISENHOWER: SOLDIER OF DEMOCRACY, Kenneth S. Davis. The full and fascinating life story of President Eisenhower up to his entrance into politics. 576 pp.

20. THE SILENT DRUM, Neil H. Swanson. Against a blazing background of bloodshed, massacre and burnings, is told the story of Frederick Van Buren, captive of the Shawnees—and his strange return to the white man's world. 552 pp.

21. EVIDENCE OF THINGS SEEN, Elizabeth Daly. Detective Henry Gamadge has to use all his skill to clear his wife of a murder charge in this fast-packed thriller. 224 pp.

22. WIND, SAND AND STARS, Antoine de Saint Exupery. The moving and spiritual story of a pioneer airman, who carried the mails in fragile planes, fought in Spain, flew the death-defying Sahara supply-lines. 256 pp.

23. OIL FOR THE LAMPS OF CHINA, Alice Tisdale Hobart. China of the 1900's—the "Open Door" years —and Americans who lived and worked there. 416 pp.

24. LONG, LONG AGO, Alexander Woollcott. Famous people, humorous anecdotes, classic crimes and caustic comment. Such famous murder cases as Snyder-Gray, Hauptmann and Hall-Mills; portraits of the famed—Katherine Cornell, Irving Berlin and George Gershwin. 288 pp.

25. CAPTAIN FROM CONNECTICUT, C. S. Forester. U. S. Navy men in the War of 1812, by a master of the sea story. The story of Captain Josiah Peabody, who fought a British convoy off Haiti, sacked and pillaged his way through the Antilles, and then was cornered by a British squadron in French Martinique. 320 pp.

26. **THE LAND GRABBER**, Peter Field. (Fight For Powder Valley!) Stevens brought blazing war to Colorado—fighting a crook who had the law on his side! 160 pp.

27. **EARTH AND HIGH HEAVEN**, Gwethalyn Graham. Against a background of religious prejudice, they fought to make a successful, happy life for themselves. 288 pp.

28. **HERE COMES A CANDLE**, Fredric Brown. Bitter, explosive story of Joe Bailey, a runner for the numbers game, a guy with a yen for cheap women and easy money. Shocking picture of fear and desire in backstreet Chicago. 256 pp.

29. **A FOREST OF EYES**, Victor Canning. Murderous intrigue and headlong passion on the edge of the Iron Curtain! He was a silent stranger, forced to protect a girl he scarcely knew from a terror he could not see. 256 pp.

30A. **MURDER ON THE LEFT BANK**, Elliot Paul. Murder in Paris; served up in a stew of dames, death and desperadoes. For careening, pell-mell excitement, nothing can match the brawling, bawdy, rip-roaring mysteries of Elliot Paul. 224 pp.

31. **THE TREES, Conrad Richter.** Pioneers in the Ohio forest by a Pulitzer Prize-winning author. Called an "unforgettable picture of the dread and horror which the earliest pioneers felt for the relentless forest." 208 pp.

32. **A MAN WITHOUT FRIENDS**, Margaret Echard. She knew he had served a jail sentence for murdering his wife. She knew that only a fluke had kept him from the gallows. Yet she put her life in his hands! 320 pp.

33. **NIGHT OF THE JABBERWOCK**, Fredric Brown. Reviewers raved about this hard-driving story of an orgy of nightmare violence that exploded one night in a town where "nothing ever happened." 192 pp.

34. **STORMY RANGE**, Dwight Bennett. A blazing, bullet-torn story of raw nerve, of a lawless country shaken by bitter hatred, and a softhearted kid who inherited a legacy of hatred. 192 pp.

35. **BRIDAL JOURNEY**, Dale Van Every. A story of brutality and horror experienced by a frontier bride captured and tortured by the savage Shawnee Indians. 352 pp.

36. **FAR FROM HOME**, Richard Mason. Compelling and powerful story of an English soldier and an Asian girl who found love in the midst of war-torn India. 320 pp.

37A. **THE SHINING MOUNTAINS**, Dale Van Every. The America of fur-trading days—the hunters and trappers who opened the West. 416 pp.

— — — USE THIS HANDY COUPON — —

BANTAM BOOKS, Dept. M2,
657 W. Chicago Ave., Chicago, Ill.

Please rush to me, postage prepaid, the selections circled below, for which I enclose $1.
(Minimum order $1. Sorry, no C.O.D.'s.)

1	2	3	4	5	6	7	8	9	10
11	12	13	14	15	16	17	18	19	20
21	22	23	24	25	26	27	28	29	30
31	32	33	34	35	36	37			

Name ....

Address ....

City.................... Zone...... State..............

10-Day *Satisfaction or Money-back Guarantee!*